Air Travel and Health

Aerospace Series List

Cooperative Path Planning of Unmanned Aerial Vehicles	Tsourdos et al.	August 2010
Unmanned Aircraft Systems: UAVS Design, Development and Deployment	Austin	April 2010
Introduction to Antenna Placement & Installations	Macnamara	April 2010
Aircraft Fuel Systems	Langton et al.	May 2009
The Global Airline Industry	Belobaba	April 2009
Computational Modelling and Simulation of Aircraft and the Environment: Volume 1 – Platform Kinematics and Synthetic Environment	Diston	April 2009
Principles of Flight Simulation	Allerton	October 2009
Handbook of Space Technology	Ley, Wittmann Hallmann	April 2009
Aircraft Fuel Systems	Langton et al.	May 2009
Aircraft Performance Theory and Practice for Pilots	Swatton	August 2008
Surrogate Modelling in Engineering Design: A Practical Guide	Forrester, Sobester, Keane	August 2008
Aircraft Systems, 3rd Edition	Moir & Seabridge	March 2008
Introduction to Aircraft Aeroelasticity and Loads	Wright & Cooper	December 2007
Stability and Control of Aircraft Systems	Langton	September 2006
Military Avionics Systems	Moir & Seabridge	February 2006
Design and Development of Aircraft Systems	Moir & Seabridge	June 2004
Aircraft Loading and Structural Layout	Howe	May 2004
Aircraft Display Systems	Jukes	December 2003
Civil Avionics Systems	Moir & Seabridge	December 2002

Air Travel and Health
A Systems Perspective

Allan Seabridge

Seabridge Systems Ltd.

Shirley Morgan

Copycat Communications Ltd.

WILEY

A John Wiley and Sons, Ltd., Publication

This edition first published 2010
© 2010, John Wiley & Sons, Ltd

Registered office
John Wiley & Sons Ltd, The Atrium, Southern Gate, Chichester, West Sussex, PO19 8SQ,
United Kingdom

For details of our global editorial offices, for customer services and for information about how
to apply for permission to reuse the copyright material in this book please see our website at
www.wiley.com.

Library of Congress Cataloguing-in-Publication Data

Seabridge, A. G. (Allan G.)
  Air travel and health : a systems perspective / Allan Seabridge, Shirley Morgan.
     p. ; cm.
  Includes bibliographical references and index.
  ISBN 978-0-470-71177-4 (cloth)
  1. Air travel–Health aspects.   2. Airlines–Employees–Health and hygiene.
3. Airplanes–Design and construction–Health aspects.   I. Morgan, Shirley.   II. Title.
  [DNLM: 1. Aerospace Medicine.   2. Aviation–instrumentation.   3. Aviation–standards.
4. Risk Factors.   5. Travel. 6. Wounds and Injuries. WD 700 S438a 2010]
  RA615.2.S43 2010
  613.6′8–dc22

                                                                              2010019291

A catalogue record for this book is available from the British Library.

Set in 10/12.5pt Palatino by Aptara Inc., New Delhi, India.
Printed in Singapore by Markono Print Media Pte Ltd

For Sue, Helen and Sarah
Allan Seabridge

For my three boys – remember that time flies
Shirley Morgan

Contents

About the Authors

Allan Seabridge I retired from the position of Chief Flight Systems Engineer at BAE Systems at Warton in Lancashire in the UK in 2006. Much of my time in the aerospace industry was spent in the area of general systems or power and mechanical systems for military aircraft, and also as team leader for the avionic systems of the Nimrod MRA 4. During my time in aerospace I developed an interest in a systems engineering approach – the understanding of the holistic aspects of systems design, including technical, process and human aspects of complex systems. Since retiring I have developed an interest in engineering education, leading to the design and delivery of systems and engineering courses at a number of UK universities at undergraduate and postgraduate level.

Personal experience always adds life to a book – even a technical one – and it has been interesting to unearth examples of how air travel might impact on health and to look back at my own flying records to consider whether I have ever been made unwell as a direct result of travelling by air.

My flying log shows that I have made over 1200 long-haul and regional flights over a period of 40 years without suffering any symptoms that I could attribute to air toxicity. I have made 43 flights in the BAE 146 type and no 'fume events' were ever observed by me personally. This is not to say that I didn't find air travel tiring, especially long haul, but I always put this down to the 'stress' of the airport experience and the long days involved in the whole trip.

Writing this book gave me good reason to monitor my recent flights (in 2009) for air quality. There has been a tendency for aircraft not to use the auxiliary power unit for cabin air while on the ramp, and when air does enter the cabin on transit to the runway there is an initial smell of burnt varnish or oil which gradually diminishes, but I can honestly say I have seen no visible contamination and have suffered no personal ill effects.

Shirley Morgan   These days I fly primarily to go on holiday, but for nine years I regularly commuted by plane up and down the UK. Like Allan, I have been fascinated by the testimonies that this book has uncovered and though I can add some experiences of my own, I think that in my case an underlying predisposition is generally the root cause of ill health, rather than air travel itself.

For example, I'm a chronic asthmatic and have twice been hospitalised with breathing difficulties within days of travelling by plane. On both occasions I was sitting next to someone who coughed and sneezed throughout the flight and, with hindsight, asking to change seats would have been a good idea, as it was probably this proximity to sick people that caused me to become ill. It could have happened at the cinema or on a bus; I believe my underlying health problem simply makes me more likely to suffer ill effects.

My youngest son has struggled with ear pressure problems since being a baby, but at 10 years old now seems to be 'growing out' of it and flies quite happily unless he is suffering from a cold, in which case the old earache and the feeling that 'his head will explode' (his words, not mine) suddenly returns.

Noise and vibration used to be my nemesis. On early-morning commutes by Jetstream 31 in the 1990s, I would almost always end the flight with a banging headache, which would invariably pass within an hour of landing. I found the aircraft noisy and could feel vibration throughout my body, and it was much worse if I sat near the window. It got to the stage where I would trample my colleagues to get an aisle seat, and I took painkillers before I boarded. Making the same flight on a BAE 146 gave me absolutely no problems, but was the Jetstream to blame, or am I a headachy person in the early morning? The jury's out on that one.

Preface

In the century that humans have been truly airborne the human body has, for the first time, been subjected to the unique physiological stresses which result from powered flight. Humans were not designed to fly. Hurtling through the atmosphere in a pressurised metal tube at incredible speed and across many time zones subjects the human body to conditions for which millions of years of ground-based evolution could in no way prepare it.

Of course flying can be fun; it's a fast and increasingly affordable way to travel. Flying can be a decades-long occupation or a one-off experience. An aircraft can be an office, a hobby or a war machine, and travelling by plane can instil feelings of weariness, fear or undiluted joy.

Unfortunately, for some people flying is a source of great anxiety, some discomfort and feelings of ill health for a period after a long flight. For a few there appear to be long-lasting effects. Some people have reported chronic illness and pain to the extent that their lives have been disrupted and professional careers ruined. These claims have been contested by both employers and the medical profession, and research is cautious and often inconclusive.

Some of this caution comes from the fact that the symptoms reported are diverse and non-specific; they also bear similarities to contested syndromes claimed by people in their ground-based lives. It is also clear that the number of people who do claim some ill effects is very small compared with the number of people who travel with no symptoms at all. But there is something there: that much is clear from the web sites that exist and the sometimes desperate messages of their contributors. This book has tried to pull together the various syndromes and effects to consolidate the information, whether it be fact, myth or the results of scientific research. We have tried to compare and contrast the conditions and considered the similar effects observed on the ground – not very scientifically, perhaps, but we aim to be informative and objective. We also hope that the information will be of use to aircraft and systems designers, to enable them to understand some of the human issues that should influence their designs. Additionally, we hope that aircraft operators can use this information to introduce the appropriate procedures to limit the exposure of their passengers and employees, and also to increase general awareness. The information could be used to influence future research and to prepare cases for funding.

Allan Seabridge and Shirley Morgan, March 2010

Series Preface

The field of aerospace is wide ranging and covers a variety of products and domains, not merely in engineering but in many supporting disciplines. This combination of disciplines enables the aerospace industry to produce perhaps the most exciting and technologically advanced products of almost any industry. A wealth of knowledge has been developed over many decades by practitioners and professionals in associated aerospace fields that is of benefit not only to other practitioners in the industry, but to those entering the industry from universities or from other fields.

The Aerospace Series aims to provide a practical and topical collection of books for engineering professionals, operators and users, as well as related professions such as commercial and legal executives in the aerospace industry. The range of topics spans the design and development, manufacture, operation and support of aircraft, as well as air transportation infrastructure operations, planning, economics, and emerging developments in research and technology. The intention is to provide a comprehensive source of relevant information that will be of interest and benefit to the wide range of professionals involved in the global aerospace industry.

Most books in the Aerospace Series typically focus on the technical aspects of Airframe, Structure and Systems - providing technical discussions that are of use primarily to engineers and designers. Other books in the Series have also covered pilot human factors and performance of the aircraft, as well as airline operations and air transport management.

This book extends the scope of the series to include the human occupants of commercial aircraft, and the impact that flying can have on their health. This has in recent years become a controversial subject, with the various health effects of air travel being hotly contested by travellers, aircraft operators and health experts. In this book the latest research on air travel and health is reviewed in an effort to pull together the sometimes disparate threads of opinion, fact and myth to establish a baseline for future discussion of symptoms and sources of research. This is a topic that will almost certainly become more visible and, very likely, even more controversial in the future, as the perceived dangers to health of air travel will become better understood.

Peter Belobaba, Jonathan Cooper, Roy Langton and Allan Seabridge

Acknowledgements

The authors would like to acknowledge the enthusiastic assistance of the following:

The ever helpful Andy Bunce of BAE Systems; Ian Moir of Moir Associates – an old and respected colleague; Wendy McCartney and Sarah Grint for their advice on helicopter issues; Shaun Hughes; John Hoyte, Chairman of the Aerotoxic Association; ex Shackleton crew members – Nev Feist (Editor of *The Growler* magazine of the Shackleton Association), Doug Cox, John Bussey, Tommy Thomas, Harry Jones for their views and experiences; Helen Seabridge for research, especially for unearthing the symptoms of the issues discussed; Caroline Roney at A.L. Simpkin & Co. Ltd, Sheffield S6 4LD, UK; Ian Taylor of Quest International UK; David Dorman of Dorway Public Relations; Andy Kanigowski, Founder of FutureFlite Corp.; Dr Peter Julu of the Breakspear Hospital; Cox & Kings for permission to reproduce their travel information leaflet.

Acronyms and Abbreviations

3D	Three-dimensional
AAIB	Air Accidents Investigation Branch (of UK Department of Transport)
A/C	Aircraft
AC	Alternating current
AEA	Aircrew equipment assembly
AEW	Airborne early warning
AGSM	Anti-g straining manoeuvre
AHWG	Aviation Health Working Group
AML	Acute myeloid leukaemia
ANC	Active noise control
ANO	Air Navigation Order
APU	Auxiliary power unit
BAe	British Aerospace (now BAE Systems)
BAE	BAE Systems
BBC	British Broadcasting Corporation
BMI	Body mass index
BS	British Standard
CAA	Civil Aviation Authority
CAP	Combat air patrol
CCFT	Close-coupled field technology

C. difficile	*Clostridium difficile*
CFC	Chloro-fluoro-carbon (compounds)
CFS	Chronic fatigue syndrome
comms	Communications (radio)
dB	Decibel
Def'n	Definition
Def-Stan	Defence Standard
DIY	Do-it-yourself
DSE	Display screen equipment
DVT	Deep vein thrombosis
EASA	European Aviation Safety Agency
EC	European Community
ECS	Environmental control system
EEA	European Environment Agency
EEG	Electro-encephalogram
EM	Electro-magnetic
EMH	Electro-magnetic health
EPA	Environmental Protection Agency
EPA	Efficient particulate (air filter)
EU	European Union
FAA	Federal Aviation Administration
FL	Flight level
ft	Foot
GA	General aviation
GCAQE	Global Cabin Air Quality Executive
GHz	Gigahertz
GP	General practitioner
H1N1	Classification of influenza virus
HAVS	Hand–arm vibration syndrome
HDL	High-density lipoprotein
HEA	Helmet equipment assembly
HEPA	High-efficiency particulate air (filter)
Hg	Mercury (chemical symbol for)
HMD	Helmet-mounted display
HOTAS	Hands on throttle and stick

HP	High-pressure (stage of engine)
h	Hour
HSE	Health and Safety Executive
Hz	Hertz

IARC	International Agency for Research on Cancer
IBS	Irritable bowel syndrome
ICNIRP	International Commission on Non-Ionizing Radiation Protection
IFALPA	International Federation of Airline Pilots Associations
IFR	Instrument flight rules
IP	Intermediate-pressure (stage of engine)
IPT	Integrated product team

| JSTARS | Joint surveillance target attack radar system |
| JSP | Joint services publication |

ME	Myalgic encephalomyelitis
MIL-STD	Military Standard (US)
mm	Millimetre
MoD	Ministry of Defence (UK)
MPA	Maritime patrol aircraft
MRI	Magnetic resonance imaging
MRSA	Methicillin-resistant staphylococcus aureus
MS	Multiple sclerosis
MSL	Mean sea level
MSOC	Molecular sieve oxygen concentrator
mSv	Millisievert

NASA	National Aeronautics and Space Administration
NFO	Naval flight officers
NRPB	National Radiological Protection Board
NRV	Non-return valve
NVG	Night vision goggles

| OBOGS | On-board oxygen generation system |
| OSA | Obstructive sleep apnoea |

| Pa | Pascal (unit of pressure) |
| Pax | Passengers |

PEC	Personal equipment connector
ppm	Parts per million
PPARC	Particle Physics and Astronomy Research Council
PRSOV	Pressure-reducing shut-off valve
PRV	Pressure-reducing valve
psi	Pounds per square inch
QC	Queen's Counsel
QFI	Qualified flying instructor
R&D	Research and development
RAF	Royal Air Force
RAM	Radar absorbent material
RAM	Random access memory
Req'ts	Requirements
RF	Radio frequency
RFP	Request For Proposal
RJ	Regional jet
RSI	Repetitive strain injury
SARS	Severe acute respiratory syndrome
SEE	Single-event effect
SOV	Shut-off valve
SPAESRANE	Solutions for the preservation of aerospace electronic systems reliability in the atmospheric neutron environment
STC	Supplemental type certificate
STFC	Science & Technologies Facilities Council
Sv	Sievert
TCP	Tricresyl phosphate
TEPC	Tissue equivalent proportional counter
UCL	University College London
UK	United Kingdom
ULPA	Ultra-low penetration air filters
US	United States
USAF	United States Air Force
UV	Ultraviolet
V	Volt
VDU	Visual display unit

VFR	Visual flight rules
VOC	Volatile organic compound
VTE	Venous thrombo-embolism
WHO	World Health Organization
WRIGHT	WHO research into Global Hazards of Air Travel
WRULD	Work-related upper limb disorder
WW2	World War II

1

Introduction

The days are long gone when people flew for the very adventure of flying, when in-flight entertainment was looking out of the window and in-flight catering was the offer of a barley sugar sweet.

Lack of sound insulation, large rotary engines and no heating provided a noisy and vibrating environment with only blankets providing some form of comfort. In the armed forces the crew spent many long and miserable hours in terrible conditions, comforted only by glucose tablets.

This age of adventurous flying ushered in travel sweets, still on sale today for the intrepid car user. A.L. Simpkin's advertising made no mistake about their use in flying:

- 'Proven to increase energy and relieve the effects of air sickness';
- '... used by RAF personnel for high altitude flying during sorties in WW2';
- '... recommended as a great way to reduce inner ear pressure during flying'.

The practice of giving sweets at take-off persisted even into the 1980s on some airlines. The theory was probably that sucking the sweet or chewing stimulated the nasal passages and ear canal to relieve any build-up of pressure. A technique known as the Vasalva manoeuvre can be used to great effect, which entails closing the mouth and pinching the nose while attempting to exhale. This counteracts the effects of water pressure in the Eustachian tubes to eliminate problems in the middle ear.

Air Travel and Health: A Systems Perspective   Allan Seabridge and Shirley Morgan
© 2010 John Wiley & Sons, Ltd

1.1 Factors Affecting Health

There are a number of factors originating in the environment of an aircraft that can have an impact on the long-term health of aircrew as a result of prolonged or habitual exposure, and which cannot be alleviated by sucking a boiled sweet. These factors may arise as a result of poor design, but more often than not they are a fact of life, a result of the physics that relates to the operation of a high-speed machine and to the environment in which it operates.

This machine can be considered as the workplace, for aircrew, in which long-term exposure and damage to health may be inevitable unless action is taken to reduce the exposure to specific hazards. In considering the aircraft in this way – as the 'office' or workplace – it is no different to the ground-based office or factory in which many humans go to work on a daily basis. Legislation exists to protect them and their employers must respect the law or suffer the consequences.

Passengers are also affected by the environment; they are after all in the same vehicle as the crew. A major difference is that their exposure is transient and irregular.

The conditions that arise from exposure to these factors may lead to problems with health either short term or, in severe cases, chronic ill health. From time to time the press covers some of the health issues of passengers or crew, especially if there is a new outbreak of ill health. The internet also carries articles on the subject and there is serious academic research being carried out to explore some issues. However, on inspection the information and the research appear to be uncoordinated – with issues discussed separately, with no single integrated viewpoint. In this book we will endeavour to bring all aspects of air travel and ill health together in one place and try to draw some conclusions.

1.2 The System of Interest

Figure 1.1 illustrates the system which is the subject of this book, showing the subsystems and their interactions.

1.2.1 The Operating Environment

The operating environment is where the aircraft spends its life – on the ground and in the air. It consists of the natural environment and the aircraft

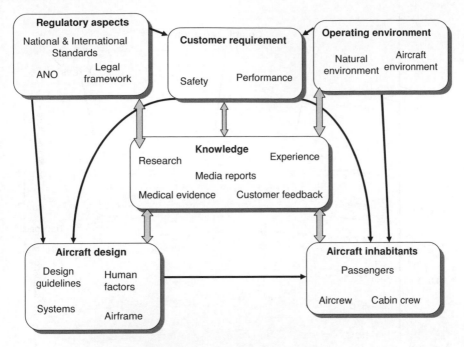

Figure 1.1 The system of interest.

environment. The natural environment is a complex interaction of chemicals and electro-magnetic radiation which forms the Earth's atmosphere. An excellent description of the atmosphere and the history of the discovery of its constituents can be found in *An Ocean of Air* (Walker, 2007). Only a small portion of the atmosphere is used by aircraft – up to about 40 000 ft (12 000 m) – although some specialised military aircraft operate above this altitude, as shown in Figure 1.2. The recent space-tourist vehicle revealed by Richard Branson is designed to operate up to 360 000 ft.

1.2.2 The Atmosphere

The atmosphere up to these altitudes is hostile to humans and the aircraft must provide some protection. The atmosphere directly affects all humans in the system – aircrew, cabin crew and passengers – but it is modified by the systems of the aircraft to provide a subtly different set of conditions. For example, the cabin conditioning system uses air from the engines to provide a comfortable cabin atmosphere for the inhabitants at the right temperature,

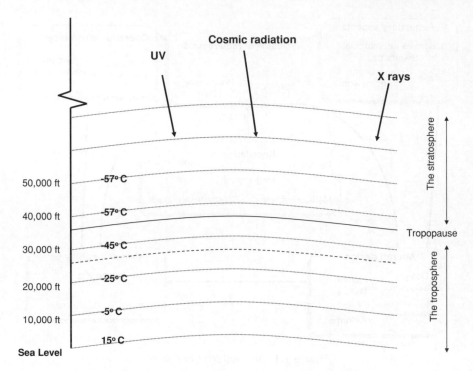

Figure 1.2 The atmosphere relevant to air travel.

pressure and humidity, whilst the engines themselves contribute to the noise and vibration experienced by the structure and the people inside.

1.2.3 The Aircraft Inhabitants

The aircraft inhabitants are taken into account in the design of the airframe and the aircraft systems. Human factors are a major element in the design of the flight deck and cabin to provide an environment that is comfortable, safe and easy to use for both crew and passengers. Guidelines developed by aircraft designers or mandated by customers are used to provide a framework for design. The aircraft design is also very much influenced by the requirements of the customer, whose specification or requirements statement will balance safety against performance, including cost and timescales. Major inputs to design are regulatory aspects such as national and international standards, legal aspects of safety, health and safety and product liability, as well as safety implications embodied in the air navigation order (ANO).

Much of the design input is influenced by knowledge emanating from research, technological advances and hands-on experience from previous projects. Medical evidence from crew health checks is used to influence procedures or design, particularly if chronic effects can be proven. At all times the media – both popular and scientific – are prepared to relate anecdotal and research-based information to the public. The internet is a fruitful source of information.

1.2.4 Sources of Environmental Stimuli

A summary of the potential sources of environmental stimuli that may lead to damage to health is illustrated in Figure 1.3.

Inside the dashed oval are the sources that are internal to the aircraft. They are generated by the aircraft or its systems or they exist as a result of the aircraft and the inhabitants. Outside the dashed oval are external sources, over which the designers and operators of aircraft have no control; these sources are natural occurrences or circumstances which must be taken into account in the design or operation of any aircraft.

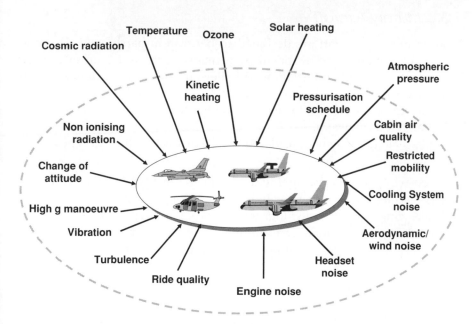

Figure 1.3 Sources of conditions that may be injurious to health.

1.3 The Aircraft

For design purposes the interior volume of an aircraft is often considered as two separate environments:

1. The airborne uninhabited environment is the space in the fuselage reserved for the installation of equipment such as avionics, for the stowage of cargo and baggage, and for the carriage of weapons or stores. This space may be ventilated, may receive warm air exhausted from the cabin, and may or may not be pressurised.
2. The airborne inhabited environment is the space reserved for occupants who require some form of life support. The space is provided with breathable air, a pressurised environment, heating and/or cooling, a means of preparing food, and suitable toilet and washing facilities.

The occupants are the subject of this book and in order to establish their position in the aircraft this section will describe who they are, where they sit and what their accommodation looks like.

1.3.1 Military Aircraft

Modern military aircraft are designed to perform in harsh conditions and operate to demanding operational performance criteria. For most of their life, though, even in an uncertain world, they rarely achieve the limits of their capability in peace-time operations. Nevertheless, the conditions encountered in flight expose the aircrew to hazards that the office worker would consider to be dangerous to health.

Military aircraft may be considered as two distinctly different types:

1. Fixed-wing or rotary-wing, highly manoeuvrable types which can exert stresses on aircrew for relatively short-duration missions.
2. Fixed-wing or rotary-wing logistic or surveillance types which may have more benign operational conditions, but fly long-duration missions and may expose crews to the same extent as commercial aircraft.

The airborne inhabited environment of some example military types is shown in Figure 1.4, together with the locations of the inhabitants.

The occupants will mostly be armed forces personnel, although there are some roles that may be provided by outsourcing to trusted civilian

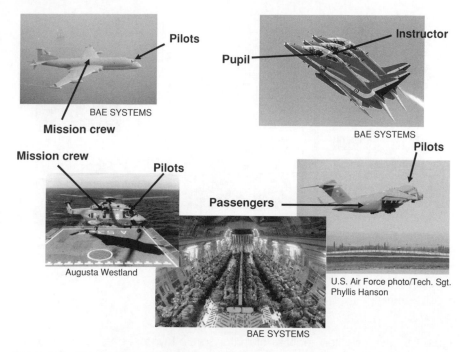

Figure 1.4 The airborne inhabited environment of example military types. Reproduced with permission from 1. BAE Systems, 2. AugustaWestland, 3. U.S. Air Force photo/Tech. Sgt. Phyllis Hanson.

agencies or government agencies such as the police, coastguard and customs and excise:

- **Pilot** – the pilot occupies the cockpit of single-seat fast jets, usually alone or in the front seat of two-seat types. Long-duration surveillance or transport types may have a first pilot and second pilot who will exchange pilot and navigator roles at regular intervals.
- **Navigator/weapons officer** – the navigator occupies the rear seat of fast jet types and doubles up as the systems and weapons officer.
- **Flight engineer** – some long-range aircraft have a flight engineer, although this role is diminishing with modern flight decks. For example, in the Nimrod MR2 the flight engineer sits behind the two pilots and monitors the performance of the aircraft systems.
- **Mission crew** – the mission crew usually occupy the cabin of surveillance aircraft, each with a role to perform and usually with a workstation.

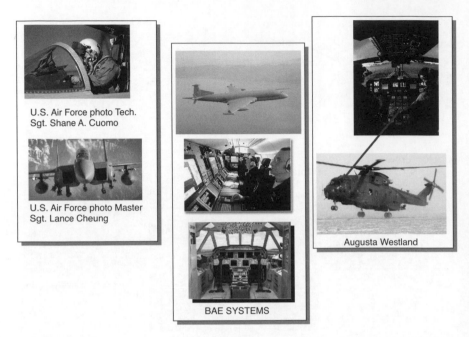

U.S. Air Force photo Tech.
Sgt. Shane A. Cuomo

U.S. Air Force photo Master
Sgt. Lance Cheung

Augusta Westland

BAE SYSTEMS

Figure 1.5 The inhabited sections of military aircraft types. Reproduced with permission from 1. BAE Systems, 2. AugustaWestland, 3. U.S. Air Force photo/Tech. Sgt. Shane A. Cuomo, 4. U.S. Air Force photo/Master Sgt. Lance Cheung.

- **Instructor** – in training aircraft tandem types the pilot instructor will sit in the rear seat, in a slightly elevated position.
- **Pupil/student** – the student pilot occupies the front seat of trainer types so that the view from the cockpit is that of the types to which the student is converting.

Examples of the interiors or inhabited sections of these types are illustrated in Figure 1.5.

1.3.2 Commercial Aircraft

Commercial aircraft generally operate in benign conditions with predictable flight envelopes and fly from large airports with facilities for servicing the aircraft and managing the passengers. For the aircrew and cabin crew the aircraft is their place of work and they may spend many hours in the cabin environment. Passengers, even frequent flyers, are not exposed to the

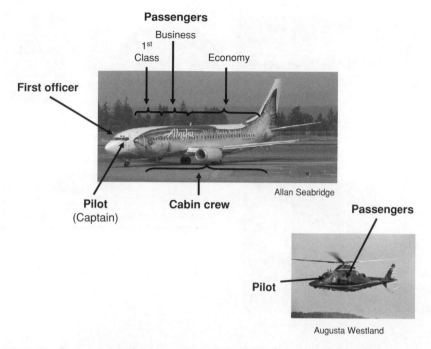

Figure 1.6 The airborne inhabited environment of example commercial types. Reproduced with permission from 1. Allan Seabridge, 2. AugustaWestland.

environment to the same extent as the crew. A description of a typical commercial airliner and crew actions is given in Midkiff, Hansman and Reynolds (2009).

The airborne inhabited environment of some example commercial types is shown in Figure 1.6, together with the locations of the inhabitants.

The occupants will mostly be a mixture of crew and passengers:

- **Pilot** – the pilot occupies the left hand seat of the flight deck and is the captain of the aircraft.
- **First officer** – the first officer occupies the right hand seat as a fully qualified pilot. The flying task is shared between the first officer and the captain.
- **Relief/check pilot** – on long-haul flights a relief pilot will be carried, who will rest and sleep in the business or first-class cabin. A jump seat between the two pilots is carried in the event that a check pilot is carried for certain mandatory pilot inspections.
- **Flight engineer** – if a flight engineer is carried, they will occupy a seat behind the two pilots.

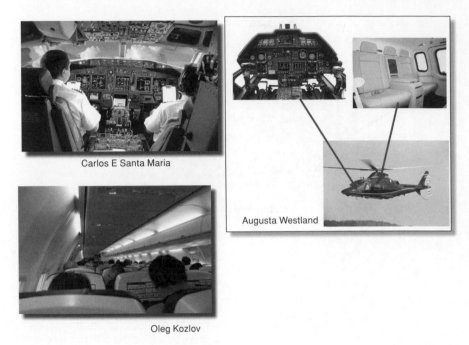

Figure 1.7 The inhabited sections of commercial aircraft types. Reproduced with permission from 1. Carlos E Santa Maria. 2. Oleg Kozlov. 3. AugustaWestland.

- **Cabin crew** – the cabin crew usually occupy the cabin with their own designated seats. Apart from providing meals and service to the passengers, the primary role of the cabin crew is to assist the passengers in the event of an evacuation or in-flight incident.
- **Passengers** – passengers are housed in cabins that are designed and equipped to a standard determined by the class of travel.

Figure 1.7 illustrates the cabin and flight deck of commercial types.

1.4 Design Considerations

The engineering teams designing these types are aiming to meet customer specifications for performance and logistic support under demanding environmental and operational conditions. Many of these conditions are also inflicted on the aircrew, together with other conditions of operation resulting from inhabiting and operating a complex military machine, most often in peace-time. Singly or in combination, these conditions can have an impact on the physical well-being of aircrew which may be apparent immediately, or may only emerge after a long period of flying. In some cases the effects may appear after flying employment has terminated.

A good systems engineer needs to be mindful of these effects and the conditions most likely to cause them. This will enable the designers of the aircraft to incorporate some alleviating aspects wherever possible, and most certainly to ensure that users of their products are aware of the risks and their duty of care to the aircrew.

'Mindful' in this case includes acquiring knowledge and experience and applying it in the engineering design of the aircraft and its systems:

- Knowledge of legislation and its impact on design and operation.
- Awareness of research in the relevant field.
- Awareness of how to merge engineering and aero-medical or physiological aspects.

Mindful may not actually be the correct word. The designer and manufacturer of military aircraft have a responsibility, a duty of care to their own test pilots and to their customers, to ensure that long-term use of the product does not jeopardise aircrew health. Therefore, there is a moral as well as a legal duty of care to users of the product. The operators of commercial aircraft have a similar duty of care to their aircrew, cabin crew and passengers.

Legislation is continuously being revised to cope with differing workplace environments to protect workers. This book describes the aircraft environment and identifies the sources of factors that can damage aircrew health and relates these to legislation – treating the aircraft as the workplace of the aircrew. Responsible manufacturers of aircraft and responsible operators do their utmost to reduce the risk, but workplace legislation often advances faster than the design lifecycle of major aircraft products, which means that there is often a difference between in-service products and legislation.

Each of the factors that can affect health in some way is discussed in terms of its impact on aircrew and consequently on working life and chronic ill health. The impact of legislation on legacy and new designs is discussed, together with mechanisms for reducing risk. Human beings are notoriously variable in their response to external factors affecting health, and many individuals do not fly for sufficiently long periods for them to represent a statistically significant sample. Nevertheless, it is worth looking for trends in aircrew and passenger comments and observations to see if there are any underlying causes.

The initial viewpoint will be that of the systems engineer, whose responsibility is to design the product to be safe to operate. This view will be complemented by a medical and human physiological viewpoint. The combination should provide guidance for engineers and contracts managers in aerospace in forming a view of the safe operation of their products and their release to service.

The book will consider the difference between commercial aircraft and military aircraft where both the operating regimes and the impact of the environmental issues are significantly different. The interpretation of legislation by commercial and military operators will also be discussed. It must be said that the commercial aircraft field has produced the main body of knowledge in the public domain, partly because of the many millions of person-flight-hours logged, partly because of the many millions of passengers, some of whom feel free to make their views known, and partly because of a lack of security which may restrict the circulation of military aircraft experience.

Aircraft provide a dynamic environment which is the daily working environment – 'the office' – for aircrew. Some aspects of this environment are particularly harsh, especially for military aircrew. Prolonged exposure to these conditions may lead to long-term damage to health unless something is done to reduce the risk. This may be by design of the aircraft and its environment or by control of flying hours. This chapter will discuss the aspects of aircraft design and operations that lead to exposure to risk, and will compare these aspects with those of the ground-based office or factory worker. It will also set the scene by describing the aircraft types and the roles that lead to the emergence of risk factors.

It is clear that there are a number of different phenomena to which aircraft inhabitants are subject, knowingly if they are employed to operate the aircraft, and unknowingly if they are paying passengers. Operators are subject to these phenomena for long periods of time, simply because they fly more often, whereas the leisure and business traveller will be only infrequently exposed. It is also clear that some inhabitants will be subjected to a number of these phenomena simultaneously, that is noise, vibration, g-manoeuvres and hard landings all in the same flight. It is, however, common to see research reports and newspaper accounts applied to individual subjects. It is important to look at the integrated system – the machine, the human and the combined effects of various phenomena.

The wise systems engineer will try to resolve this issue by taking into account all aspects of design of the vehicle. It should be noted, however, that organisations often divide their engineering teams into functional responsibilities and that makes it difficult to take an integrated viewpoint.

The staff of a company designing and releasing to service an aircraft need to be aware of the implications of the impact of their design on the inhabitants of the aircraft. All staff should be aware of legislation. The company should publish procedures and processes that ensure that engineers are given guidance on where to look for standards, how to apply them and how to deal with any deviations. Training should be made available to ensure that engineering staff are fully briefed on contemporary legislation.

Tracking and understanding legislation is an essential task for all aircraft companies. The requirements of both immediate and prospective customers must be included in the design standards. The requirements of certification agencies and government agencies mandating on health and safety must also be understood. Figure 1.8 illustrates an example process for monitoring and applying legislation and making a contribution to a body of knowledge that can be used for future updates of regulations.

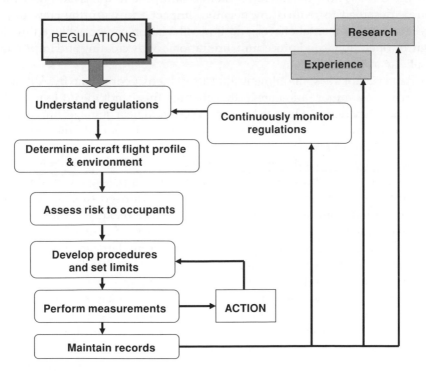

Figure 1.8 An example process for monitoring and applying legislation.

1.5 Summary

So it seems that short-term discomfort is almost inevitable. What one does not expect is long-term or even chronic illness. Yet there is evidence that some aircrew, cabin crew and passengers do suffer. Those who do fly for work should expect their employers to make some provision for reducing the risk, in response to various health and safety legislation. Operators have a duty of care to passengers to reduce their risk.

The following chapters will present a picture of the potential hazards to health that are posed, together with mitigating action that can be provided by aircraft design by the application of suitable operating procedures.

We also examine the issues and their effects upon aircraft inhabitants and explore the current body of research, debate and anecdotal evidence. It is important that aircraft and system designers are aware of the facts, in order to ensure that their design takes into account the real issues of their customers. We also discuss how the aircraft systems contribute to the issues, how they can be designed to reduce any negative impact and warn the crew of the presence of detrimental effects, and how alleviation can be achieved by design, by operational and procedural limitations, or by creating and improving public awareness.

Chapter 2 looks at some ailments that affect many people who fly. Although referred to as 'minor' ailments in this chapter, this is not meant to trivialise the issue. They may not lead to long-term physiological damage, but they are of real concern to those who suffer and for whom the very thought of flying causes great concern.

Chapter 3 examines the quality of air provided to fast-jet military pilots as oxygen-enriched pressurised air, and the air in a pressure cabin of large military types and commercial aircraft with pressurised cabins. Generally air is extracted from the engine and conditioned to make a comfortable environment for passengers. There have been reports of contamination from the bleed air causing discomfort and even long-term damage to health. This chapter will examine the symptoms and some potential technical solutions including one new and innovative piece of technology.

Chapter 4 looks at deep vein thrombosis (DVT), which is a controversial subject affecting both crew and passengers. Although there is no proven link to air travel, the subject does enjoy a wide press and many in-flight magazines include a section on exercises to perform during flight to reduce the risk of potential damage. The chapter examines aircraft cabin arrangements and seating to look for causes and solutions.

Chapter 5 considers the sources and effects of noise and vibration – separately and combined. Comparisons are drawn with other workplaces in which noise and vibration are major issues.

Chapter 6 deals with exposure to cosmic radiation as well as non-ionising radiation from radio frequency sources such as radar transmitters. Ionising radiation from cosmic rays is a fact of life, made slightly more of a threat to humans by long exposure at high altitude.

Chapter 7 considers the occurrence of circumstances which can lead to back and neck injury, excluding direct impact such as use of an ejection seat. There are aspects of seating for normal use that can lead to chronic back pain.

Chapter 8 examines some issues specific to military fast-jet crews. This includes the effects of acceleration as an issue for military crews, particularly those on fast jets. The impact of sustained high-g manoeuvres on immediate performance and on long-term health will be examined, as well as the impact of heavier helmets and the possibility of neck/vertebra damage. Heat stress is another issue – the military pilot is exposed to a wide range of temperatures, including direct solar radiation in high-flying fast jets. In many cases the pilot wears clothing to ensure survival in the event of an accident and also rubber garments designed to reduce the impact of high-g manoeuvres. Combined with a high metabolic rate in a stressful combat or training scenario, this may lead to heat stress. Pressure oxygen breathing is used by military fast-jet crews and there is potential for hypoxia as a result of loss of oxygen or reduced oxygen concentration caused by cabin pressurisation failures.

Chapter 9 examines the situation where aircrew are expected to use a workstation to perform their task throughout the mission. This may be aircrew who make use of cockpit displays and controls throughout the flight to control the aircraft and its route, or it may be mission crew on military surveillance aircraft who must analyse sensor data for long periods of time using screens and keyboards in the cabin.

Chapter 10 looks at legislation in the UK and Europe that is aimed at health and safety in the workplace as it exists at the time of writing. Legislation is continually being reviewed and revised and readers are expected to familiarise themselves with the latest situation.

Chapter 11 describes a generic process that is used by aircraft designers to ensure there is a rigorous process for including in the aircraft design those requirements that determine the performance and safety of the product and ensuring that all aspects of legislation are considered. This process continues throughout the in-service life of the aircraft to ensure that comments received during operations with the public and with users are suitably recognised.

Chapter 12 will examine the information in the previous chapters and try to establish some common causes and solutions that are known. The impact of the effect of one or more of the issues will be discussed as an integrated system view.

Each chapters ends with a view of what can be done in the system to provide some form of alleviation. This includes not only engineering solutions in the aircraft systems, but also solutions in the wider system of interest illustrated in Figure 1.1. The solutions may be technical, such as the introduction of control of the systems or the introduction of warnings, or may be procedural and influence the manner in which the aircraft is operated or the crew deployed.

References

Midkiff, A.H., Hansman, R.J. and Reynolds, T.G. (2009) Airline flight operations, in *The Global Airline Industry* (eds P. Belobaba, A. Odoni and C. Barnhart), John Wiley & Sons, Ltd.

Walker, G. (2007) *An Ocean of Air: A Natural History of the Atmosphere*, Bloomsbury.

Further Reading

Belobaba, P., Odoni, A. and Barnhart, C. (eds) (2009) *The Global Airline Industry*, John Wiley & Sons, Ltd.

Useful Web Sites

publications.parliament.uk/pa/ld200708/ldselect/ldsctech

2

Some 'Minor' Ailments

2.1 Introduction

This chapter looks at some ailments that affect many people who fly. Although referred to as 'minor' ailments in the title, this is not meant to trivialise the issue. They may not lead to long-term physiological damage, but they are of real concern to those who suffer, and for whom the very thought of flying causes great concern. Often air travel is associated with a long journey which includes car or rail travel, security checks, queues, buying tickets, arranging local transport and spending longer than usual in an atmosphere of stress. Even without a genuine fear of flying, most people are troubled by the anticipation of this 'hassle' together with a background worry about accidents or terrorist incidents.

2.1.1 Health Information

The airlines provide awareness on the subject of well-being and health issues for passengers, either on their web sites or in their in-flight magazines. As well as giving essential safety information, there is usually a page in the magazine that outlines the following:

- Reassuring passengers' anxiety about safety and crew professionalism.
- How to relax during the flight.
- The dangers of excessive alcohol consumption.
- Reducing dehydration.
- Coping with ear and nasal discomfort.

Air Travel and Health: A Systems Perspective   Allan Seabridge and Shirley Morgan
© 2010 John Wiley & Sons, Ltd

- Reducing jet lag.
- How to exercise whilst seated.

2.1.2 Considering the Whole Trip

When considering the factors that lead to passenger complaints, or for travellers to resent their trip, it is important to consider the whole event of air travel and not to concentrate simply on a single flight. Any trip that involves air travel also includes a number of other events, and it may well be factors in the entire trip that contribute to any form of ill health. Consider a typical business trip:

- Getting up at an unfamiliar hour.
- Travelling to the airport – by car, train, bus or a combination of these.
- Time spent at the airport – checking in, security, waiting in the lounge.
- The outbound flight – maybe with a stop-over or inter-terminal transfer(s).
- Jet lag, travel to hotel, business meetings, entertainment.
- Hotel – unfamiliar sleeping conditions, late night, disrupted sleep.
- Travel to the airport – by car, train, bus or a combination of these.
- Time spent at the airport – checking in, security, waiting in the lounge.
- The return flight – maybe a stop-over or inter-terminal transfer(s).
- Waiting for baggage.
- Travelling home.

To many people this is an unusual circumstance, since their normal day is not like this at all. For most people it happens only at holiday times. The business traveller, on the other hand, may be subject to this trip on a weekly basis. No matter how many times the frequent flyer does it, it never quite becomes a routine or happy experience.

2.1.3 Some Symptoms Reported by Travellers

An article in the *Telegraph* (Starmer-Smith, 2009) lists a range of symptoms experienced by a number of crew members since they started flying, which include:

- High blood pressure
- High cholesterol
- Anaemia
- Pneumonia/bronchitis

- Asthma
- Infertility
- Chronic fatigue
- Insomnia
- Depression
- Eczema/psoriasis
- IBS/Crohn's disease
- Fibromyalgia
- Thyroid disorder
- Breast cancer
- Osteoporosis
- Chronic sinusitis
- Tachycardia
- Peripheral nerve damage
- Obstructed lungs
- Osteoarthritis.

These symptoms have been classified by some general practitioners as 'aero-toxic syndrome', see Chapter 3. Many people will not experience any of these, let alone all of them, but these symptoms and others are experienced by some travellers; the following chapters will examine them and they will be discussed in Chapter 12.

2.1.4 Health Risks

Business passengers may be protected by their employer, but leisure passengers fly at their own risk. In the event that some systemic cause is found for a flying-related issue, then recompense may be sought from the airline operator or the aircraft designers. As discussion on the internet shows, this is no easy matter.

In the UK the House of Lords Select Committee on Science and Technology considered the matter in 2000 and issued a report of its findings and recommendations. The report acknowledged the significant growth in air travel as well as the developments in medical knowledge which have established links between environmental factors and health. The report also acknowledged the right of regulators and the aircraft industry to give absolute priority to safety; however, it criticised them for failing to give sufficient attention to health and environmental factors. One of its proposals was that there should be improvements in regulatory arrangements, and that more information should be made available to potential air travellers.

2.2 Some Common Conditions of Air Travel

2.2.1 General Discomfort

There are many people who really enjoy air travel – it is their route to freedom, friends, family, holiday, adventure. They treat it as part of the journey, no more tedious or dangerous than the car, bus or train. To others, however, it is a source of anxiety and discomfort. Even in this modern age, there are still people who have not flown and do not relish the prospect, no matter how hard some airlines try to make it an enjoyable experience.

Even to people accustomed to flying, especially long haul, there are some discomforts. Delays, queues, inter-terminal transit lugging cases, security checks, unsuitable food at unsuitable times – these all lead to feelings of frustration and annoyance. The accompanying noise, vibration, air temperatures and different forms of transport, the effect of cabin altitude on ears, nose and digestive system – all seem like part of a prolonged attack on the body's physiology.

One of the authors, undertaking the long trip from home in the UK to a hotel in Fort Worth, Texas, just wanted the journey to end. Early-morning travel by car from home to Manchester Airport, Manchester to Chicago, long walk to change terminals, Chicago to Fort Worth, bus to rental car centre, navigate to Fort Worth, check in, unpack – even then it was still early evening and not time for bed. And this from an experienced long-haul traveller who thought he knew how to maintain an eating and sleeping routine!

There are a number of discomforts that air travellers put up with. To some people they are a minor and occasional irritant. To others they are a severe and debilitating trial that they endure each time they travel. This can ruin the travel experience, and in some cases cause severe anxiety that has an impact on personal life.

The whole trip, as described above, does not always make for a joyous experience: 'I didn't care where I was, though perhaps I would have preferred to be on the moon . . . but was still in that state of passive acceptance of life or death . . . which is the only way to survive the stupefying effects of long-haul jet travel' as the author Jenny Diski observed after a trip to New Zealand (Diski, 2006).

An indication of the trauma of the whole experience can be gauged from this report from the *Guardian* (Sherine, 2009):

> Imagine, after frantically shoving the contents of your bedroom into your suitcase, terrified that you've forgotten something vital such as tickets, passport or child, you rush to the airport. Running late, you're surrounded

by passengers who are manically trying to catch planes themselves, ranting at staff or staring miserably at the floor in boredom, as endless PA announcements echo around the soulless waiting areas.

Then you're told your liquid containers are too large and will have to be junked, the online voucher you printed was the booking confirmation and not the proper ticket, your luggage is too heavy and you owe them all your travel cash, and it's the last call for your flight. And up next: Naked Stranger Machine! It's enough to get anyone knocking back the duty free bleach.

In addition to all this, there is the disruption to life, a change of routine, food and drink, anxiety, business pressures and the prevailing state of health of the traveller. Holiday travel is much the same sequence of events but including exposure, often overexposure, to the Sun, 'foreign' food, too much alcohol and the pressures of supervising children. It is possible that the lifestyle accompanying air travel makes a significant contribution to the overall perception of the trip.

2.2.2 Jet Lag (Circadian Dysrhythmia)

One of the major causes of illness amongst air passengers and crew is jet lag, primarily caused by travelling across time zones. Frequent flyers often adopt a specific routine that helps them avoid the fatigue and other physical symptoms, such as headache, insomnia and gastric upset, that jet lag can bring.

Tiredness can be a big problem for aircrew and research suggests that performance can be impaired as a result. Although the primary cause of accidents is usually human factors, jet lag and sleepiness are rarely officially identified as the reason for inadequate performance in the aircraft. Incidents have been recorded in which fatigue and jet lag on long-distance flights have been identified as a background factor (Myagi, 2005). Research on how to combat these has looked at strategies such as on-board napping, crew augmentation and light therapy (Samel, Wegmann and Vejvoda, 2009).

Some flight crew say they simply 'get used' to body clock disruptions and airline schedules should be designed to give crew sufficient rest periods on the ground. Even so, crew members may find it difficult to adjust to time changes and start their next shift without being adequately rested. Some female flight crew complain of menstrual irregularities as a result of frequent travel across time zones. A survey carried out amongst New Zealand flight attendants regularly flying long haul revealed that 96% suffered jet lag symptoms, most usually tiredness, lack of energy and a susceptibility to colds and throat problems.

A report for the House of Lords Select Committee on Science and Technology (2000) in the UK recommended that jet lag should be studied as a compounding effect of deep vein thrombosis (DVT) as part of Phase II of the WRIGHT Project. It also called for the Civil Aviation Authority (CAA), as the body responsible for the health and safety of aircrew, to commission a study into the possible long-term health effects that jet lag might have.

Travelling any distance by air can be tiring, but avoiding caffeine and alcohol, staying well hydrated and resisting the temptation to sleep to pass the time are the best defences against jet lag on a long flight.

Studies suggest that it takes one day for each time zone crossed to recover from jet lag. The most generally accepted 'cure' on arrival is to get onto 'local' time as quickly as possible – so even if you land at midday after a 14-hour flight, having hardly slept, you should resist the temptation to go to bed until it goes dark. However, if you are staying for only a day or so, as is the case with most crew layovers, it may be better to keep close to your sleep pattern at home.

This is underlined in a study by Lowden and Akerstedt (1998) which tested the idea of letting aircrew retain their home-base sleep/wake pattern during layover. Instead of adopting local sleep hours, 19 flight attendants were scheduled to a westward layover (50 h) flight (Copenhagen–Los Angeles, 9 h) on two occasions.

On one trip, crews adopted the local sleep pattern and on the other trip retained home-base sleep hours. They were monitored for 10 days before, during and after the flight and kept sleep/wake diaries.

Ratings of jet lag symptoms and sleepiness were greatly reduced during layover, but not at home, for the home-base condition. It was also found that jet lag feelings seemed to be mainly 'sleepiness' and the number of awakenings during sleep. It was concluded that retaining the home-base sleep pattern may reduce jet lag during layover.

Various remedies are available to combat jet lag. The drug melatonin is sometimes used by frequent travellers, as the hormone helps to synchronise the body, but studies on its benefits are inconclusive. It is not generally available in the UK.

An anti-jet-lag diet, developed by Chicago University, works by using meals to reset the body clock, speeding up or slowing down the metabolism. It has been tested on National Guard troops flying across time zones.

Other 'revivers' include homoeopathic remedies with ingredients such as arnica and camomile; eye masks and 'sun' lamps to simulate light and dark; and even 'thermostimulation' where heated pads are strapped to the body and electric currents passed through them. The aim is to improve circulation and release endorphins, to improve well-being and aid sleep.

One of the authors uses a simple technique when flying back from the US: have a shower in the Admiral's Club lounge, when on-board ask for two large Jack Daniels, ask the cabin staff not to serve the meal after take-off, put on the noise-cancelling headset and sleep until breakfast time. It seems to work.

2.2.3 Fear of Flying (Aerophobia)

The old joke 'I'm not afraid of flying, I'm afraid of crashing' does actually sum up the fear of flying for most people, and, yes, sufferers do accept that the probability of dying in an air accident is significantly less than dying in a car crash on the way to the airport. But in reality there are many different phobias that can contribute to a fear of air travel, such as claustrophobia (a fear of enclosed spaces) or acrophobia (a fear of high places). Some people simply do not like the feeling of being 'out of control' whilst others suffer motion sickness and are panicky about vomiting during a flight. A prior event, perhaps a 'bad' landing years earlier, can also turn into extreme anxiety, and because symptoms can start suddenly after years of previously worry-free flight, it can be disastrous for people whose job requires air travel. In this case air travel is not a cause of ill health, but it can be a stimulus for conditions that bring considerable stress.

Flight phobias and fears cover such a spectrum and can have so many root causes that they cannot simply be categorised as ill health related to air travel. There is little any aircraft designer or manufacturer can realistically do to calm the fears of every anxious passenger.

The good news for those who suffer is that cures are possible, with treatments ranging from hypnotherapy to training courses and self-help books. Also available are courses, often operated by airlines such as Virgin Atlantic and British Airways, that desensitise passengers to the fear by counselling and in-flight experience.

In some cases of panic attack the symptoms include nausea, vomiting, constipation or diarrhoea, dry mouth, dizziness, lack of concentration, temporary inability to concentrate, shivering, sweating, numbness and palpitations. These symptoms may recur on one or more occasions (fearfreeflying.co.uk).

2.2.4 Discomfort of the Ears

Perhaps the most common minor ailment is the effect of changes in pressurisation that cause passengers to feel pain in the ear. This can be exacerbated if passengers have a cold; in fact passengers with colds or nasal problems are advised not to fly until the illness is over. This is often impossible for business travellers whose flights are scheduled for specific meetings, whilst those people on holiday may have limitations on their ticket use and are often forced to fly on fixed dates.

Passengers are advised to adopt one or more methods of reducing the problem; the in-flight magazine usually provides information on suitable remedies. These include swallowing or pinching the nose, swallowing or blowing out against a closed mouth – even chewing gum can help to relieve the symptoms. In the worst case damage to the ear drum can result, with young children and babies at particular risk, being unable to 'equalise' the pressure in their ears. Ear pain is the reason why many infants cry on take-off and descent.

Pressure changes can adversely affect the middle ear, sinuses, teeth and gastrointestinal tract. Any sinus block (barosinusitis) or occlusions that inhibit equalisation of external pressure with pressure within the ear usually result in severe pain. In extreme cases, rupture of the tympanic membrane may occur. Maxillary sinusitis may produce pain that is improperly perceived as

Figure 2.1   Children and babies are at risk (Surkov Vladimir). Reproduced with permission from Shutterstock Images. See Plate 4 for the colour figure.

a toothache (an example of referred pain) and pain related to trapped gas in the tooth itself (barondontalgia) may also occur.

Ear block (barotitis media) also causes loss of hearing acuity (the ability to hear sounds across a broad range of pitch and volume). Pilots and passengers may use the Valsalva manoeuvre (closing the mouth and pinching the nose while attempting to exhale) to counteract the effects of water pressure on the Eustachian tubes and to eliminate pressure problems associated with the middle ear. When subjected to pressure, the tubes may collapse or fail to open unless pressurised. Eustachian tubes connect the corresponding left and right middle ears to the back of the nose and throat, and function to allow the equalisation of pressure in the middle ear air cavity with the outside (ambient) air pressure. The degree of Eustachian tube pressurisation can be roughly regulated by the intensity of abdominal, thoracic, neck and mouth muscular contractions used to increase pressure in the closed airway.

Young children and babies can suffer pain on take-off and landing , caused by an inability to equalise pressure in the ears. It is difficult, if not impossible, to instruct a baby or very young child (Figure 2.1) how to perform alleviation techniques, although older children can be assisted or taught.

References

Diski, J. (2006) *On Trying to Keep Still*, Little, Brown.

House of Lords Select Committee on Science and Technology, 5th Report (November 2000) Air Travel and Health, Chapter 1: Summary and Recommendations.

Lowden, A. and Akerstedt, T. (1998) Retaining sleep and wake patterns in aircrew on a 2-day layover on westward long distance flights. *Chronobiology International*, **15** (4), 365–376.

Miyagi, M. (2005) *Serious Accidents and Human Factors*, John Wiley & Sons, Ltd.

Samel, A., Wegmann, H.M. and Vejvoda, M. (2009) *DLR-Institute of Aerospace Medicine*, Linder Höhe, Germany.

Sherine, A. (2009) A scanner darkly. *Guardian*, 15 October.

Starmer-Smith, C. (2009) Illness among cabin crew heightens toxic air fears. *Telegraph Travel*, 18 July.

Further Reading

Belobaba, P., Odoni, A. and Barnhart, C. (eds) (2009) *The Global Airline Industry*, John Wiley & Sons, Ltd.

Jonathan, D.A. (2002) Otolaryngology, in *Aviation Medicine and the Airline Passenger* (eds A. Cummin and A. Nicholson), Arnold.

Nicholson, A.N. (2002) Sleep disturbance and jet lag, in *Aviation Medicine and the Airline Passenger* (eds A. Cummin and A. Nicholson), Arnold.

Stott, J.R. (2002) Airsickness, in *Aviation Medicine and the Airline Passenger* (eds A. Cummin and A. Nicholson), Arnold.

Useful Web Sites

af.mil
antijetlagdiet.com
fearfreeflying.co.uk
nojetlag.com
publications.parliament.uk/pa/ld200708/ldselect/ldsctech

3

Air Quality

What would you do if, whilst sitting at your desk, you developed a pounding headache which quickly turned to nausea, difficulty in breathing and an overwhelming desire to go to sleep? You would probably pack up for the day and go home. But what if your 'desk' was the flight deck of an airliner, and you were part way through an eight-hour shift?

Illness can strike anyone, at any time, but a concern for those who fly for a living is that the air they are breathing actually makes them ill. A pilot or flight engineer feeling sleepy and mildly confused sounds like a recipe for disaster, and there is a growing log of incidents – both medically documented and anecdotal – which suggest that some people are adversely affected by the air that they breathe in flight, with calls for more research into how the quality of breathable air in aircraft can impact on the health of both crew and passengers.

More than a decade ago, the US aviation industry asked the American Society of Heating, Refrigerating and Air-Conditioning Engineers to form a panel to investigate air quality inside airliners. In the UK and Australia it has also been on industrial and government agendas and is particularly relevant during health crises like the H1N1 swine flu pandemic, when there are public fears about the transmission of potentially lethal viruses.

Does in-flight air make people sick? If so, what are the problem elements, who is at risk and what systems can be put in place to prevent it?

Air Travel and Health: A Systems Perspective   Allan Seabridge and Shirley Morgan
© 2010 John Wiley & Sons, Ltd

3.1 The Environment

3.1.1 The Atmosphere

Aircraft operate at altitudes from mean sea level (MSL) to above 50 000 ft, spending long durations in flight and also on the ground being serviced, cleaned and boarded by passengers. The external environment varies considerably and this has an impact on the internal environment in which the crew and passengers are located.

Air is essential to human life and it is becoming clear that air quality is being affected by pollution from industry and fossil fuel combustion products. Already in big cities and industrial conurbations air quality is of concern, to the extent that it features in weather forecasts. The external atmosphere relevant to aircraft is shown in Chapter 1, Figure 1.2. At lower altitudes there is plentiful oxygen in the air and plenty of biological processes in place to remove carbon dioxide and produce oxygen. There is, however, lots of pollution from human endeavour. At higher altitudes the low partial pressure of oxygen makes it less available. People accustomed to living at high altitudes have adjusted, and some very fit climbers have succeeded in climbing the highest mountains without additional oxygen assistance, but without some kind of acclimatisation any physical effort becomes difficult, and the immediate effect of exposure can result in headaches, shortness of breath, fatigue, nausea and pain from expanding air trapped in tooth fillings.

3.1.2 Air Quality

A perceived general reduction in air quality has led to concerns about a rise in respiratory illness in the general population. People are becoming sensitised to pollutants such as pollen, carbon particulates and animal or human skin flakes and hair. This has resulted in more awareness of air pollution and its symptoms.

This awareness and its reflection in regulatory documents have led the public to expect good air quality, whether in their cars or in public transport, and this includes aircraft for business or leisure travel. As a result air passengers now expect not just comfortable seats, entertainment and plentiful food and drink, but a comfortable 'breathing' environment. This makes them happy and more likely to pay the airline to fly with them again. In the case of military passengers or mission crew, they have a job to do. This entails looking at the information provided by on-board sensors for long periods of time and making decisions that may affect the battlefield or military intelligence, so they need to be comfortable in order to be alert and vigilant.

Accordingly, the designers of aircraft environmental control systems (ECSs) strive to provide an environment with:

- a plentiful supply of draught-free clean air;
- an appropriate range of temperatures;
- the right level of humidity.

3.1.3 Cabin Air Supply

Air is usually obtained from the aircraft engines to pressurise the cabin. Although it is conditioned, there is still the possibility that it contains contaminants such as oil vapour, anti-icing fluid or exhaust from ground vehicles. These fluids and gases are likely to be ingested from the ramp or taxiway environment into the engine intake and hence into the bleed air. Combustion products may arise from contact with hot compressor stages of the engine. This may have a detrimental effect on the respiratory tract, particularly for aircrew and passengers with temporary or permanent respiratory ailments.

The quality of air in the cockpit and provided as oxygen-enriched breathing gas is largely determined by the filtration in the ECS and the air distribution system. There is a risk that contaminants in the atmosphere will not be filtered successfully, or may even be present in the system pipework as a result of cleaning fluid residue or the presence of toxic material entering the system, for example cadmium. There may be long term-cumulative effects as a result. Figure 3.1 shows a block diagram of an example air system and the potential sources of contamination marked with an arrow.

3.1.4 Sources of Contamination

The engine compressor seals prevent contaminated air being fed into the air system, but there is a possibility that these seals will lose effectiveness during their life. Ozone is removed from the bleed air at a certain altitude to prevent it irritating the passengers' bronchial systems. The flow of air in the cabin is arranged to avoid cross-contamination between passengers, and a filter is included in the recirculation line to remove dust and biological contamination. The APU seals should prevent bleed air being contaminated. The APU, however, operates on the ground, usually in a busy airport environment, and its intake air will be polluted by other aircraft exhaust, ground vehicle exhausts and vapours from spilled fluids on the tarmac. It may also be subject

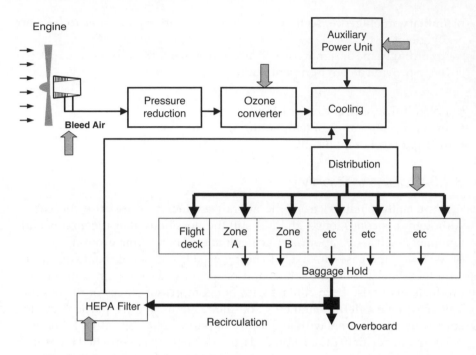

Figure 3.1   An example aircraft air system and sources of contamination.

to reingestion of air from its own exhaust or from the aircraft engine exhaust depending on the ramp environmental conditions, such as wind direction or heat rising from the tarmac.

The emissions from engine combustion are known and their impact on health as a result of contamination of the aircraft external environment is recognised (Marais and Waitz, 2009). A table in Marais and Waitz lists some impacts on health of various chemical species, which bear a striking resemblance to the main symptoms listed below:

- **Volatile organic compounds (VOCs)** – temporary effects such as nausea, fatigue and throat irritation.
- **Sulphur dioxide** – breathing affected, aggravation to existing respiratory and cardiovascular disease.
- **Nitrogen oxides** – lung irritation, bronchitis, pneumonia, lower resistance to respiratory infections.
- **Carbon monoxide** – breathlessness, impaired vision, manual dexterity, learning ability, performance of complex tasks, threat to sufferers of cardiovascular disease, angina.

- **Particulate matter** – aggravates existing respiratory and cardiovascular diseases, effects on respiratory system, produces alterations in the body's defence systems against foreign materials.
- **Ozone** – damages lung tissue, reduces lung function, affects healthy adults and children.

It should be noted that VOCs are present in many domestic situations. House cleaning compounds such as polishes and cleaning fluids contain VOCs and have been the cause of discomfort after use. In a recent TV programme, one house owner described their effects as 'tiring, debilitating'. VOCs are also said to be the cause of 'new-car smell', a combination of synthetic materials in seats and panels, as well as adhesives. This is also said to cause some people discomfort. VOCs are also present in plug-in electrical deodorisers and scented candles and are said to be a trigger for asthma (BBC News, 25 August 2004).

3.2 Aircraft Environments

Different aircraft types operate at different altitudes and have different flight profiles. An indication of the regions of the environment at which each type cruises is shown in Figure 3.2.

Aircraft operating below 10 000 to 12 000 ft do not generally have pressurised cabins, which includes light aircraft for leisure flying and general aviation (GA) types used for business travel. Regional aircraft and long-haul aircraft are usually pressurised. Military types are usually pressurised: fast jets because of their need for rapid changes of altitude and reconnaissance, transport or tankers because they operate at high altitudes.

3.2.1 Commercial Aircraft

The commercial aircraft cabin is designed to house a number of passengers and crew for the duration of the flight in comfort. That means that the cabin 'altitude' is ideally equivalent to a pressure that is never more than that of 8000 ft and is more often set lower than this. The air circulation meets the requirements for an appropriate number of changes of air per minute and moisture content. A typical cabin pressurisation schedule is shown in Figure 3.3.

The number of passengers can range from about 10 for a business jet, 40 to 100 or so for a regional aircraft and many hundreds for a transoceanic

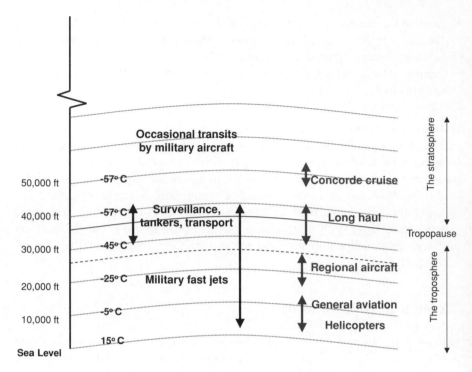

Figure 3.2 Cruise altitudes for various aircraft types.

passenger jet. Cabin volume varies considerably over the range of commercial aircraft. Taking the Airbus range alone, cabins in the different types are designed to suit passenger loads from 107 to over 400.

The needs of the occupants differ too, as does their perception of the cabin environment. Passengers have paid to be on the flight, they may be travelling for leisure or for business and they want to be comfortable during what, for some, is still a traumatic experience. They will eat, drink and rest or sleep and some may work. Oxygen will be provided in the event of depressurisation for the relatively short period of time that the aircraft needs to descend to below to about 10 000 ft. Passengers will notice any deficiencies in the cabin environment, be it too hot, too cold or draughty, and they may well complain.

The cabin crew will be working. They must stay alert to the passengers' needs, and need to be ready and fit if an emergency arises, but unlike paying passengers they will tend not to complain about cabin discomfort; it is their working environment after all.

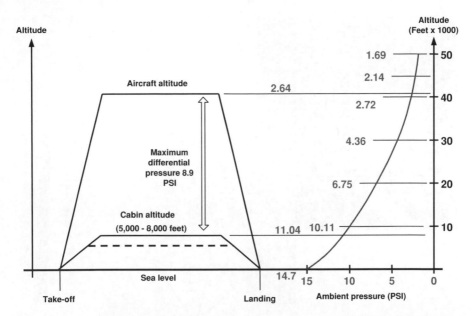

Figure 3.3 A typical commercial aircraft cabin pressurisation schedule. Reproduced with permission from Moir & Seabridge (2008).

The aircrew need to concentrate on flying the aircraft, staying awake through the routine procedures and long flights, unless taking a designated rest period, and they will be provided with their own personal supply of gaseous oxygen for emergencies.

3.2.1.1 Changes of Air

Most modern aircraft provide the cabin with air from an air-conditioning system over which the passengers have no control. Previous systems, many still in use, allowed passengers to direct air over their bodies using a 'punkah louvre' or overhead variable air inlet device above their seats. This panel is shown in Figure 3.4. The temperature and flow of air are now controlled by the cabin crew, and pressurisation or cabin altitude is controlled by the aircrew.

3.2.2 Military Aircraft

A fast jet will experience many rapid changes of altitude during training or combat missions, and will often operate at high altitudes in direct sunlight. The cockpit volume of fast jets and trainers is small, but the cockpit receives

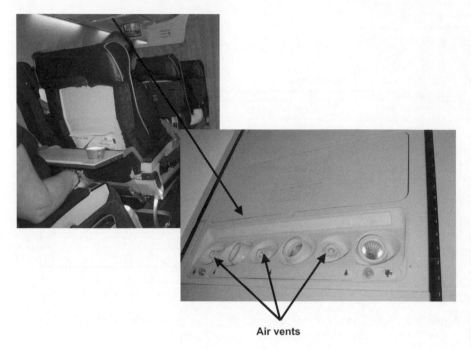

Air vents

Figure 3.4 Overhead panel showing air control nozzle (Allan Seabridge). Reproduced with permission from Allan Seabridge.

a large heat load from a combination of direct solar radiation, kinetic heating, the pilot's own metabolism and the avionic equipment packed into the consoles and main instrument panel. This can approach a heat load of 8 kW, which requires a lot of cold air to maintain a comfortable cockpit temperature. As well as being blown into the cockpit generally, cold air is sprayed at the body and face.

The pilot and navigator/weapons officer/pupil will be wearing masks and will normally breathe pure oxygen or an oxygen–air mixture under pressure. Modern aircraft use bleed air from the engines fed to an on-board oxygen generation system (OBOGS) that enriches the air with oxygen by removing nitrogen in a bed of zeolite (Moir and Seabridge, 2008).

This is shown in Figure 3.5, in this case for a two-seat aircraft, although the architecture is the same for a single-seat aircraft, but with only one regulator. The following description is from a Honeywell Aerospace Yeovil paper (Yeoell and Kneebone, 2003):

Engine bleed air enters the Pre-Conditioning system element where the temperature is reduced, ideally to less than 70°C and water is removed as

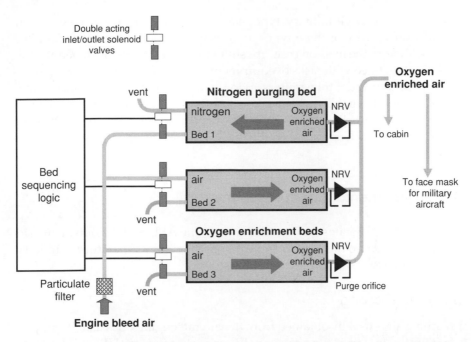

Figure 3.5 On-board oxygen generation system (Honeywell Aerospace Yeovil) (Moir and Seabridge, 2008; Yeoell and Kneebone, 2003). After Lawrence & Kneebone (2003). See Plate 5 for the colour figure.

far as possible. In addition it is normal at this stage to use a combined particulate and coalescing filter to remove potential contaminants including free-water that may still be contained in the inlet air.

The OBOGS contains a pressure reducing valve to reduce the inlet air pressure of the air supply to that required by the OBOG Generator, typically 35 psig.

The next system element is the Oxygen Generator, or more correctly, the OBOGS Concentrator that uses multiple zeolite beds to produce the oxygen-rich product gas.

The system monitor/controller is a solid state electronic device that monitors the PPO_2 level of the OBOGS concentrator product gas, and adjusts the cycling of the beds to produce the desired level of oxygen concentration for cockpit altitudes below 15 000 ft. This process is known as concentration control and means that no air-mix, or dilution, of the product gas is required at the regulator, hence preventing the ingress of any smoke or fumes from the cabin into the pilot's breathing gas supply.

The cabin of large military types is very similar to that of commercial aircraft, since many of these types are designed around a commercial platform. Although the mission crew are small in size, up to about 18 people, the cabin is often filled with racks of equipment. Reconnaissance types, for example, will contain radar transmitters, avionics equipment, radios, workstations and a galley. The aircraft will also fly a profile similar to passenger-carrying aircraft – take-off, climb, cruise on station, return cruise, descent and land. Modern types, such as the Nimrod MRA4, will be equipped with OBOGS as described above.

3.3 Environmental Control Systems

The aircraft ECS obtains air from the engine, the APU and the outside world and distributes it to the cockpit and cabin at the right temperature, humidity and cleanliness to support the occupants.

3.3.1 Air Cooling

Breathing air is usually obtained from the aircraft engines. It is usually tapped off from the intermediate-pressure (IP) or high-pressure (HP) stages of the engine where it is very hot, and then passed through heat exchangers and refrigeration systems before being used to pressurise and condition the cockpit or cabin. Bleed air is used for a number of tasks on the aircraft. Figure 3.6 shows the bleed air provided by a number of sources – engine, APU or ground supply – and its bleed points for anti-icing and pressurisation. Bleed air for the ECS and pressurisation is fed to the cabin.

A typical cooling arrangement is shown in Figure 3.7 where bleed air at high temperature from the engine is reduced to a suitable temperature to be fed to the air distribution system.

The aircraft may well be parked for long periods of time without the engines running, for example when boarding or disembarking passengers at an airport, and whilst waiting to obtain clearance to start engines. During this period air is bled from an APU and fed into the cooling and distribution system to provide comfortable cabin conditions.

3.3.2 Air Distribution

A typical air distribution system in shown in Figure 3.8.

To enable the cabin conditions to be regulated, the cabin is divided into sections and temperature control is provided for each section by the cabin crew.

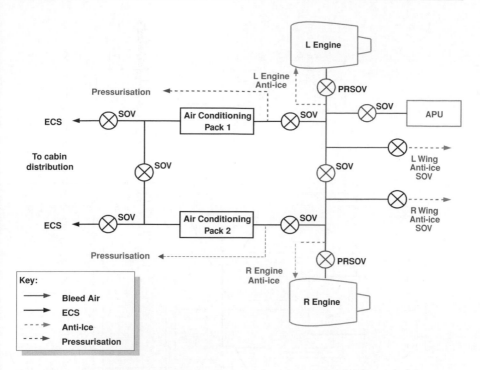

Figure 3.6 Sources and uses of bleed air on the aircraft. Reproduced with permission from Moir & Seabridge (2008).

Example zones are illustrated in Figure 3.8 which shows the temperature control system schematic. Air enters the cabin at roof level and is extracted at floor level, flowing downwards to avoid cross-contamination.

A modern passenger transport aircraft ECS recirculates up to half of the cabin air. Whilst the ECS fresh air is treated to remove ozone, the recycled cabin air is filtered for many unwanted contaminants, including:

- Micro-organisms (moulds, bacteria and viruses)
- Dust (including fibres and skin flakes)
- Odours
- VOCs (including bio-effluent from food and drink, oil, fuel, hydraulic fluid and ice protection fluids).

Such filtration is achieved using a high-efficiency particulate air filter (HEPA) conforming to European Standard 1822-1. This standard defines efficiencies

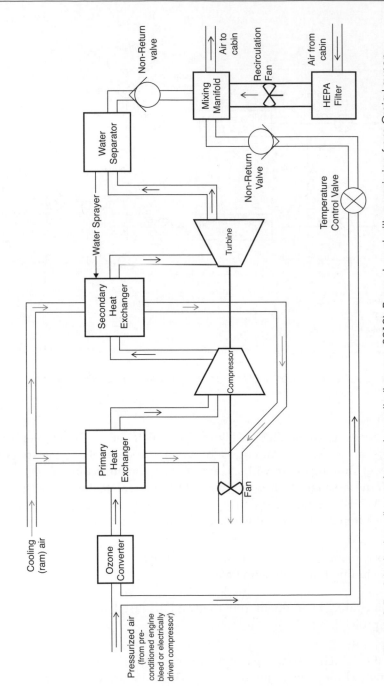

Figure 3.7 A typical cooling system schematic (Lawson, 2010). Reproduced with permission from Craig Lawson.

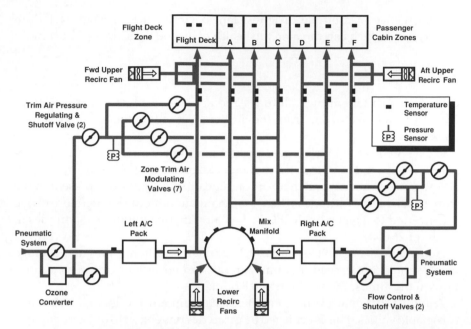

Figure 3.8 A typical air distribution system schematic. Reproduced with permission from Ian Moir.

from 85 to 99.995%, and filters on modern transport aircraft perform at the upper end of this scale with efficiencies of at least 99.95%.

Bacteria (0.5 to 1.5 microns) and viruses (0.01 to 0.1 microns) are extremely small. However, a typical HEPA filter on a modern aircraft has the performance to ensure that the recirculated air has a micro-bacterial content as low as the fresh outside air.

HEPA air filter systems are available that are specially designed for aircraft use and contain anti-microbial additives to protect against a range of bacteria, fungi and yeast. The filter system by itself can trap submicron particles and the use of anti-microbial additives ensures that any potentially disease-causing substances are destroyed on contact. The addition of an anti-microbial compound to the standard HEPA filter system is necessary because micro-organisms such as bacteria, fungi, yeast, mould and mildew are able to reproduce and grow even in the dry environment of a passenger aircraft. The anti-microbial compound penetrates and disrupts the cell walls of these organisms, making them unable to reproduce.

Without these special HEPA filters, biological contaminants can grow and recirculate throughout the aircraft. This is especially alarming because of the emergence of multiple drug-resistant bacteria such as Mycobacterium tuberculosis which can threaten the health of airline passengers and crew.

3.3.2.1 Ozone

Ozone is an irritant at high altitudes and an ozone converter is used above 26 000 ft to remove ozone from the air before it enters the air-conditioning pack.

Ozone (O3) is a colourless gas, most of which present in the Earth's atmosphere is suspended in the stratosphere. However, ozone is an irritant, even at low concentrations, and there are significant concentrations present in the troposphere at the altitudes where long-haul aircraft cruise. Therefore ozone is converted to oxygen by the ECS when aircraft are operating at altitudes in excess of 26 000 ft.

Acceptable ozone levels peak at 0.25 parts per million (ppm) and average levels should not exceed 0.1 ppm. Aircraft can encounter ozone concentrations in the atmosphere as high as 0.8 ppm.

A fully functional ozone converter will maintain levels well below the acceptable values. This is achieved by catalytic conversion to oxygen, and further dissociation occurs upstream in the (main engine or electrically driven) compressor and downstream in the air-conditioning packs when the ozone comes into contact with components.

A catalytic ozone converter is a pressurised (by bleed or ram air) vessel containing a palladium or palladium and metal oxide catalyst. The catalyst is arranged in a fine honeycomb mesh to maximise the surface area that the air passes over, thereby ensuring effective operation at air high mass flow rates and low temperatures.

Despite the reaction being catalytic, the ozone converter will need to be maintained since contamination and corrosion of the honeycomb core may occur over time.

3.4 Health Issues

Whilst most passengers in commercial aircraft will be familiar with the environment and relaxed, others will be anxious and may not relish the experience. Most will experience some discomfort from the effects of pressurisation as well as the changes in altitude, which, combined with their prevailing state of health, may cause temporary physiological changes. Most passengers do not fly sufficient hours to expect any long-term chronic problems, but crews, either flight or cabin, do fly frequently and may be experience long-term exposure to flight conditions. Some of the issues that arise are discussed below.

3.4.1 Effects of Contamination

There have been incidents of smoke and fumes in the cabin air as a result of oil or hydraulic fluid entering the engine bleed air because of leakages in the lubrication systems, or as a result of debris left in the air distribution ducting during servicing.

In passenger aircraft the air is recirculated for economy reasons, passing through filters to remove biological contamination sources. However, it has been found that engine anti-wear additives in jet oil can lead to contamination of the cabin air (Starmer-Smith, 2009). Although it is conditioned, there is still the possibility that it contains contaminants such as oil vapour or combustion products. This may have a detrimental effect on the respiratory tract, particularly for aircrew with temporary or permanent respiratory ailments. Recent research on samples of cabin residue found high levels of tricresyl phosphate (TCP). Exposure to this can lead to symptoms of drowsiness, respiratory problems and neurological illness. Passengers often complain of 'feeling unwell' after long-duration flights, and recent experience shows that aircrew are suffering illness that has been said to lead to early cessation of flying duties.

3.4.2 Aerotoxic Syndrome

The group of symptoms and illnesses reported by aircrew and passengers has been classified as aerotoxic syndrome, and the Aerotoxic Association has created a web site (aerotoxic.org) to bring together sufferers. The syndrome, sometimes called sick aircraft syndrome, is defined as 'a severely debilitating condition that can affect air passengers and air crew after they have been exposed to noxious fumes from the jet engines entering the cabin. This is known as a "fume event".'

Statistics suggest that 1 flight in 2000 suffers some form of fume event, but the problem may be widely under-reported due to ignorance or as a result of commercial pressure on pilots and maintenance personnel. Another complication is that most fumes have no discernable smell and the definitions of a fume event and the distinction between smoke, fumes or haze and odours are sometimes unclear (aerotoxic.org).

Ironically, it can be argued that the move to banish smoking from the skies has contributed to the problem of air toxicity by reducing the need to refresh cabin air more frequently.

The Aerotoxic Association's web site contains descriptions of the experiences of victims. This sample is necessarily unrepresentative of the total population for the following reasons:

- The sample size is very small compared with the total number of passengers, aircrew and cabin crew who fly regularly.
- The victims are those who associate their symptoms with the syndrome represented by the site and who have sufficient incentive to write in.
- It only includes those people who have access to the internet.
- It only includes those people who have complained; that is, those who are aware of the issue.

Nevertheless the results are interesting in that they describe a condition that a number of people feel they have suffered from to such an extent that it has changed their life. Some report debilitating symptoms and a premature end to their flying careers. Others have been forced to reduce their flying hours and have suffered financial difficulties as a result. Many complain of incomplete diagnoses and the uncertainty of wondering what is actually making them ill. A few suspected they had developed psychological problems.

The most common symptoms reported are given in Table 3.1 and classified in groups of pilots, cabin crew and passengers (Pax). The most common symptoms are briefly described below:

- **Headache** – variously described as pressure in the head and pain in the temple; this symptom was experienced by nearly all subjects.
- **Breathless/tight chest** – trouble with breathing or shortness of breath, with some pain in the upper respiratory tract and chest tightness.
- **Tiredness/fatigue/exhaustion** – numerous mentions of feeling tired, fatigued or exhausted with loss of energy and extreme lethargy.
- **Light-headed/giddy** – sometimes with loss of balance.
- **Memory/word recall** – short-term memory loss, difficulty with remembering the right word or completing a sentence.
- **Vision impaired** – blurring, sore eyes; irritation, burning sensation.
- **Concentration** – losing the ability to concentrate on routine tasks.
- **Dry/sore throat** – coughing, throat infections.
- **Nausea** – feeling sick, nauseous.

Table 3.1   List of reported symptoms.

Subject	Pilots								Cabin crew			Pax		
	2	3	5	6	8	9	10	11	1	4	7	12	13	14
Symptom														
Headache		X	X	X	X		X		X	X	X	X	X	X
Breathless/tight chest	X	X			X			X	X	X	X	X	X	
Tired/fatigue	X	X	X	X	X	X	X							
Light headed/giddy	X	X				X						X		
Poor memory/word recall	X	X	X		X	X			X	X	X	X		
Vision impaired	X	X				X								
Lack of concentration		X			X	X				X			X	
Dry/sore throat		X	X		X				X			X		
Nausea					X	X			X			X		X
Depression		X					X			X				
Low motivation	X													
Flu symptoms			X						X					X
Night sweats														
Metallic taste										X				
Palpitations										X				
Muscle spasm/shake	X								X	X				
Tingling sensation				X	X					X				
Hand/feet swelling										X				
Coughing					X					X				
Wheezing										X				
Digestive problems									X	X				
Drowsiness			X											
Balance impaired							X				X			
Airway irritation					X						X			
Eye irritation					X									
Rash/burning	X				X								X	
Detached mood	X						X							
Insomnia			X		X						X			
Diarrhoea	X	X								X				X

Despite the range of symptoms listed, not all victims reported a diagnosis. In some instances tests were carried out and no diagnosis was given, but the wide-ranging diagnoses that were provided, in some instances more than one per victim, are listed below:

- Pleuro-pericarditis
- Toxic encephalopathy
- Lupus
- MS
- Dementia
- Fibromyalgia
- Depression
- Stress
- Epilepsy
- Alzheimer's disease
- Hypo-gonadism
- Low white cell count
- Super low HDL (high-density lipoprotein) cholesterol
- Brain damage from chemical exposure.

3.4.2.1 Diagnosis

Marais and Waitz (2009) discuss the impact of engine combustion products on the environment and the potential impact on health, whilst stating:

> emissions are known to have a direct impact on human sickness and lead to an increased risk of premature death. In addition the link between emissions and climate change … is becoming ever clearer with continuing research. … Emissions impacts occur over several different timescales. Air quality is immediately affected and varies on a daily basis with emissions volumes, while health impacts may take longer to emerge and tend to persist for longer periods.

Whilst this is valid for wide-scale environmental impact, the cabin environment is a closed environment, albeit only for the duration of any one flight, and emission impacts are likely to be radically different.

The pilots, cabin crew and passengers reporting air toxic episodes not only suffer from symptoms ranging from respiratory difficulties to exhaustion, headaches and muscle tremors, but also have the problem of medical misdiagnosis.

Because they often present 'flu-like' aches and pains, sometimes with nausea, the symptoms are attributed to colds or influenza, even jet lag, and it is only when the symptoms become long term that other avenues may be investigated.

Clinicians can face many difficulties when attempting to diagnose. Many of the problems reported by aircrew and passengers are indicative of other medical conditions. For example, a small-scale study has been done (Starmer-Smith, 2009b) comparing the symptoms of MS with those reported as a result of air toxicity. There are some striking parallels between the two sets of symptoms, but samples studied so far are too small for conclusions to be drawn.

An investigation in 2009 revealed that one in seven of 789 UK airline staff questioned had to take more than a month's sick leave in the previous year. They had a higher than normal incidence of cancer and 1 in 20 had been diagnosed with chronic fatigue syndrome. The study was carried out in conjunction with the British Airlines Stewards and Stewardesses Association, which has called for a full official investigation (Starmer-Smith, 2009b).

In the same month, an American Airlines attendant launched a lawsuit against Boeing, claiming that she had been made ill by toxic fumes on-board an aircraft. A lawsuit brought against Alaskan Airlines by 26 flight attendants who claimed they had been made ill by toxic leaks was successful.

Some aircrew report such a wide range of symptoms that it is hard for a precise diagnosis to be reached; many say they simply feel poisoned. Others eventually 'give up' on conventional medicine and treat themselves with complicated detox regimes. One retired airline pilot actually bought himself a book on Alzheimer's disease, so sure was he that his catalogue of long-term problems could only be attributed to a gradual decline in brain function.

3.4.2.2 Impact on Life and Work

There are many stories of careers curtailed by illness attributed to air toxicity in aircraft, but feeling ill is not the only reason that some pilots chose to quit – some say they simply feel unfit to fly, and are worried that problems with concentration and memory are putting their colleagues and passengers in danger. People who fly for a living can feel pressured to carry on despite the problems, and have mortgages to pay and often love their careers, but some are hurt by implications (often by employers) that their illness is imagined and resolve to carry on no matter what. Others say they have been threatened with termination of their contracts whilst on sick leave.

Writing on the Aerotoxic Association's web site, one pilot with over 30 years' experience said it was the reaction of friends and family that finally persuaded him that something was very wrong. He said:

> My fatigue became chronic and eventually I could not complete my simulator check successfully – yes, I did actually fall asleep in the simulator at one point! Following this I set about a series of medical tests to identify the problem. These covered all the normal checks plus EEG (body) and MRI (head) scans, blood tests and neuropsychometric examination. Everything appeared normal except the last which showed a huge discrepancy between ability and performance i.e. Short term memory loss.

There are other reports of memory impairment. One pilot describes himself as a shadow of his former self. He said:

> I still train (on doctor's advice) but I am no longer the marathon-running, rugby-playing training captain I once was. I have virtually no short term memory. Every thing I do, I do from a list which I carry around with me. I have the same conversations over and over again. Strange things also happen. I get lost in a town I know well. Often I get in the car and then don't know where I am going or why.
>
> (*Source*: aerotoxic.com)

It can take months, or even years, of tests before the term aerotoxic syndrome is finally added to a sufferer's medical records. Some never get what they consider to be a satisfactory diagnosis.

3.4.2.3 Medical Advice

Passengers can find advice on what to do to avoid 'fume events' and subsequent ill effects, but for those working on aircraft for long periods, an avoidance strategy is harder to adopt.

Advice for passengers includes sitting near the front, the logic being that in many aircraft the air passes from front to back, so the freshest, most oxygen-rich air is found in front of the wings. Directing the overhead air vent away from the face is also suggested, because what passengers may think is 'fresh' air, of course is not.

The third strategy is to ask cabin crew for the cabin ventilation to be increased. The mix of fresh and recirculated air in the cabin can sometimes be adjusted if several passengers request the change.

Advice is given to passengers on the subject of travel fatigue (British Airways web site):

Travel fatigue is different from jetlag. It's a combination of the stress of travelling and the sleep debt you accrue whilst travelling. If you're suffering from travel fatigue, your ability to function properly can be impaired:

- judgement and decision making can be reduced by 50%,
- communication can be reduced by 30%,
- memory can be reduced by 20%,
- attention can be reduced by 75%.

It is interesting to note that these symptoms are similar to those experienced by victims of aerotoxic syndrome.

3.4.3 At Risk on the Ground?

Long-haul passengers frequently complain of illness during or immediately after a flight. One common thread is that many people fall sick when their aircraft has been forced to wait on the ground for a period of time. Air quality on aircraft is often worse when it is on the ground because air circulation systems sometimes are not operated while passengers board, or when aircraft have to sit for long ground delays.

There have been calls in the UK for a 30-minute limit on the amount of time passengers should remain in an aircraft when the ventilation systems are non-operational to help prevent the spread of infectious diseases. A House of Lords report said risks increased significantly when passengers are closely confined in the aircraft with no effective filtration and called for a reporting system to be put in place to record those instances when a 30-minute period is exceeded in order to put pressure on airlines to do everything possible to avoid such events.

3.4.4 Spreading Illness

Global alarm at the spread of the SARS virus saw the introduction of airport screening for passengers. The rapid spread of the virus had, in part, been caused by infected people travelling from one part of the world to another by air. Figures showed, however, that the risk of in-flight transmission of SARS was very small. Only a handful of probable cases were reported worldwide, and all those occurred before airport screening procedures were put in place.

The way in which viruses such as SARS are spread underlines the view of many experts that, when diseases are spread on-board aircraft, they are simply passed by proximity, not by air circulation systems. Passengers coughing and sneezing can easily infect others nearby, but viruses and bacteria are not spread through the entire aircraft. In other words, you are just as likely to become infected if you sit next to a sick person on a busy train.

3.5 System Implications

Systems designers need to maintain an awareness of the implications of their system on passengers and crew in a technical environment in which changes in one system may have a severe impact on another. Hence changes to engine oil additives may have produced contamination which is not being filtered out by conventional filters. Technological advances need also to be monitored and included as necessary in design updates.

3.5.1 Contaminants

New contaminants contained in the engine air as a result of new compounds used in the engine lubricant system are believed to be a problem as they enter the cabin. Work is currently being performed to sample the air in the cabin of aircraft to identify the compound, but only when this is completed will it be possible to design a filter to remove the contamination from cabin air.

Work is also being conducted in the UK by Cranfield University to track 'emission incidents'. Once the frequency of such incidents is understood, then work will be undertaken to identify contaminants and measure their concentration.

The same group of chemicals blamed for air toxicity in flight is also held responsible for Gulf War syndrome and illness caused to farmers using sheep dip (organophosphate pesticide). Of course these chemicals do not affect everyone, but those who are susceptible can suffer devastating health problems and it is hard to predict who could be most at risk.

A 2004 study by Dr Sarah Mackenzie-Ross (2006) of UCL estimated that up to 197 000 passengers may have been exposed to fumes on-board aircraft in that year, on UK flights alone. Another study (Mackenzie-Ross, 2008) looked at the cognitive function of 27 airline pilots who had reported exposure to contaminated air. It concluded that neuropsychological and physical defects reported by the pilots resembled those found in cases of exposure to organophosphates and solvents and recommended further research.

Dr Peter Julu is a consultant neurophysiologist who believes there should be a large-scale study, carried out with the co-operation of the major airlines, to investigate the link between impaired cognitive function and air toxicity. He has advised the House of Lords Science and Technology Group and since 1995 has seen 25 commercial airline pilots on referral and retired 5 on medical grounds. All had presented with neurological problems. Initially he suspected carbon monoxide poisoning, but comparisons with a group of his patients known to have suffered it in a mining incident threw up some important differences. There was, however, a striking similarity with a group of patients who had worked as farmers, abattoir assistants or livestock inspectors and who had been exposed to organophosphates in sheep dip. He said:

> The pilots referred to me all had more than seven years experience and all reported a fume event, usually pre-flight. Their symptoms were a mirror image of the farmers. Impaired cognitive function such as memory loss leads to obvious dangers for an airline pilot, and this can be exacerbated by rank. At least one patient admitted he had simply forgotten to do some of his pre-flight checks, but his more junior co-pilot did not feel able to overrule him. Some patients also report periods of disorientation during flight.
>
> The pilots I have seen come from many countries, but all have significant experience and love their job. Some have been accused of malingering; most had no idea what was happening to them. I think many pilots ignore the early symptoms, or just put them down to tiredness.
>
> I believe a large scale study sponsored by major operators is the only answer. My own samples are too small to draw scientific conclusions, but it seems to me that if something is poisoning them we should find the definite cause so that we can prevent it and offer the correct treatment to those affected.
>
> *(Author interview)*

A BBC *Panorama* investigation in 2008 showed how passengers could be exposed to toxic fumes in the cabin. Swabs and air samples taken during two UK flights allegedly showed traces of the neurotoxin TCP. The programme looked at claims from pilots and passengers who said they had become ill as a result of breathing in toxic fumes. Even though the contamination levels were well within international safety standards, there were calls for better control and more studies into 'fume events' in flight (BBC News, April 2008).

In the US a panel of aviation industry experts has recommended voluntary standards for on-board air circulation, lower ozone exposure, new monitoring for contaminated air from oil or hydraulic fluid leaks, and limits on pesticides used on aircraft. The *Wall Street Journal* (July 2009) reported that the Association of Flight Attendants was pressing for better monitoring, saying that these issues should have been addressed 'a long time ago'.

A document has been provided to health care providers as a guide for examining patients and providing an explanation of their symptoms. The guide provides a summary of its aims:

> Outside air is bled off the engines/auxiliary power unit and supplied to cabin/flight deck on commercial aircraft. Under certain conditions, toxicants such as pylorized engine oils and hydraulic fluids may leak into the aircraft cabin and flight deck air supply systems. Airline workers may develop acute or chronic health effects and seek attention from health care providers. This quick guide focuses on oil exposures.
>
> *(Source:* available at ohrca.org/healthguide.html)

The Global Cabin Air Quality Executive (GCAQE), which represents aviation workers on air quality issues, has been critical of a government investigation into contaminated air. It said that although the study was welcome, the research methods used were 'inadequate' and called for a public inquiry (Daily Mail, March 2008).

A study for the Australian Department of Defence (2008) looked at organophosphate and amine contamination of cockpit air in the BAE Hawk, the F-111 and Hercules C-130 aircraft. It too has called for more research into the incidence and implication of air toxicity. Commissioned after concerns raised about contaminated air in the BAE 146 were widely publicised and linked to passenger and aircrew health problems, the report was produced for the Ministry's Defence Science and Technology organisation. It looked at the Hawk and the F-111 because of historic incidents of 'smoke' in the cockpit and at the C-130 because of reports of odours from bleed air, possibly from engine oil or from amine additives used as anti-oxidants.

Air samples from the Hawk were collected from the aircraft on the ground with the APU running. Of 15 samples taken, the highest had a TCP concentration of $49\,\mu g/m^3$, but most samples showed much lower readings, typically of less than $1.5\,\mu g$. In the F-111 and C-130 samples were taken from both ground engine runs and in flight. All the samples were found to be under the maximum permissible levels. One of the 'high' readings taken in the Hawk was found to be associated with an oil spill near the engine, but as the sample was taken with the cockpit open, TCP concentrations could have actually exceeded acceptable levels.

The report recommended that total TCP air concentrations of less than $1 \mu g / m^3$ should be desirable, rather than have a statutory exposure limit of $100 \mu g$. This was because of uncertainty in some of the data and lack of information on economic constraints and the potential impact on aircrews. It said those targets appeared to be achievable, based on the sampling, and added that the low levels of toxicity 'are indicative of the satisfactory condition of the compressor oil seals'.

In addition to TCPs, trialkyl phosphates showed up in the cockpit samples taken from the F-111 and C-130. These were attributed to hydraulic fluid contamination but were considered to be 'of low toxicity'.

Further research was recommended, along with measures such as washing the ECS heat exchangers with a solvent during maintenance, which could reduce toxicity levels further. Interestingly, the report highlighted that flight crews may have become sensitised to odours in the aircraft. During the study odours were detected by the crew without being apparent to two of the report authors, who were also present.

The report also touched on the question of who is actually at most risk from toxic air. As the two 'high' TCP samples taken from the Hawk were when it was on the ground, the question arises of whether it is maintenance mechanics, rather than flight crew, who are more vulnerable to exposure.

It is worth noting that the use of OBOGS on modern aircraft may itself pose a risk of air pollution. If it is proven that engine bleed air is contaminated then consideration must be given to the filters and beds of OBOGS to determine if the contamination can be removed. If not, then it is being fed under pressure into the pilot's mask.

If the contamination is removed in OBOGS, then consideration must be given to the risk posed to the supplier's staff tasked with cleaning, drying or recharging the OBOGS zeolite beds to ensure that residual contamination is not a health and safety risk.

3.5.2 Future Systems

Aircraft such as the Boeing 787 are leading the field in designing conditioning systems that use fan-pressurised ram air rather than engine bleed air to charge environmental conditioning systems – this is essentially free of contaminants from the engine. The benefits to air quality are secondary, the main impetus for this type of system stemming from a desire to free the engine of external bleeds to provide consistent operating conditions. Nevertheless it does provide air for the majority of the aircraft flight free from any form of combustion products. This does not mean that the APU, if

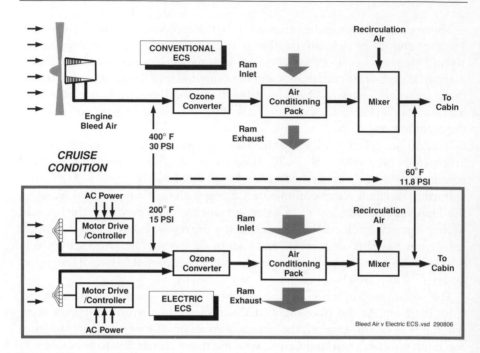

Figure 3.9 Comparison of bleed air and electric ECS (Ian Moir). Reproduced with permission from Ian Moir.

used to provide conditioning on the ground, will not provide contaminated bleed air. This system is illustrated in Figure 3.9 and compared with a conventional system.

The conventional system is shown drawing air from the engines at high pressure and high temperature. In comparison the 787 obtains its air from the external environment at relatively low pressure and temperature, the air being compressed by electrically driven turbines prior to entering the cooling cycle.

3.5.2.1 Air Manager

BAE Systems and Quest International UK have introduced a radical new active cabin air management system to the world's airline industry. Named AirManager, it is said to eliminate all airborne viruses and bacteria and will set a new standard for clean air on-board passenger and cargo aircraft.

Through its Regional Aircraft business, BAE Systems has engineered and obtained full European Aviation Safety Agency certification for the installation of the new system on its BAE 146/Avro RJ airliners. The company has also signed an agreement with Quest International UK to act as authorised distributor for worldwide AirManager sales to airlines for an initial five-year period. Trials involving Boeing 757 airliners are already underway after BAE Systems designed and secured a supplemental type certificate (STC) for use on this aircraft. STC designs for other aircraft types will be determined in line with market demand, with initial efforts expected to focus on the Airbus A320 and Boeing 737 airliners.

Passive HEPA filters are dependent on air being dry for maximum efficiency. With condensation or humidity the filter will become wet and can freeze, resulting in back pressure and flow restriction problems. Current filter technology is significantly restricted in further improved performance to meet both today's and future expectations, for example anti-icing fluid and engine or APU exhaust ingestion, and the heightened public awareness of airborne viruses (common cold virus, swine/bird flu, SARS, etc.).

The patented AirManager uses close coupled field technology (CCFT) – a contained and safe electric field that eliminates smells and breaks down and destroys airborne pathogens, contaminants and toxins.

Quest R&D into non-thermal plasmas in the late 1990s revealed that this form of energy could be easily harnessed and adapted into a very efficient method of sterilising air. The first commercial application of CCFT was in nursing homes to remove odours, which naturally led to protecting patients compromised by harmful viruses and biohazards (*C. difficile*, MRSA, etc.). The technology was also applied to problems caused by VOCs in manufacturing operations such as in pharmaceuticals, ink production and solvent use. This extensive experience in the medical and health care fields has provided a sound foundation for the transfer of technology to aerospace applications.

AirManager works by using the non-thermal plasma field of CCFT, which has a significant impact on the widest possible range of biological and chemical air quality challenges. Existing plasma fields are generated by using a high-voltage/high-power source which is allowed to discharge between two electrodes. The resultant ionisation process reacts with the surrounding environment with positive and negative results. The patented CCFT uses a high-voltage coil/dielectric arrangement to generate an electron avalanche which effectively breaks down long-chain molecules, much like a mass spectrometer would. Any resulting residues are then trapped by electrostatic media which have been designed to be disposed of in an environmentally friendly manner.

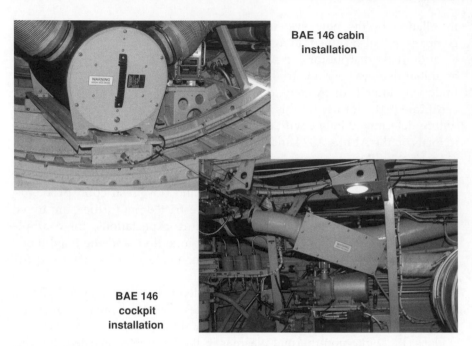

BAE 146 cabin installation

BAE 146 cockpit installation

Figure 3.10 AirManager in the BAE 146 aircraft (BAE Systems). Reproduced with permission from BAE Systems.

CCFT treats any airflow by actively sterilising the air. It generates an electron avalanche which has a totally disruptive effect on all biohazards, VOCs and chemicals passing through it. Viral particulates, bacteria, yeasts, moulds and spores are destroyed, whilst effectively reducing VOCs and other chemical compounds in a single pass. The electrostatic media provide high airflow/minimal pressure drop capability whilst arresting any ultimate particulates to below 0.1 micron, outperforming HEPA standards (certified Diffused Oil Particulate testing), guaranteeing the delivery of a new standard in cabin air quality.

CCFT is very low powered (3.6 W) and thus has minimal impact on aircraft power usage with a mass between 3 and 5 kg depending on the installation. Installation in the BAE 146 and Boeing 757 aircraft is shown in Figures 3.10 and 3.11. The installation consists of one close coupled field and three electrostatic filter elements and a transformer mounted on the outside of the housing (no airflow restriction). The unit is made from aluminium-welded sections. One unit is required aft of each engine bleed air input; the Avro RJ requires two units and the passenger 757 requires five units whilst the 757 freighter version requires one unit.

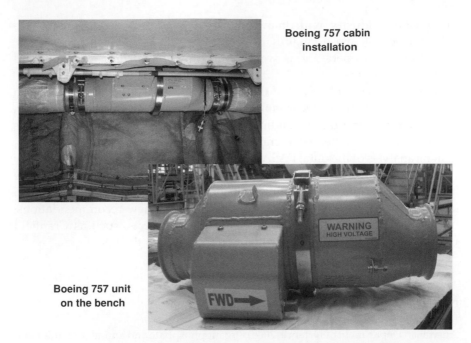

Boeing 757 cabin
installation

Boeing 757 unit
on the bench

Figure 3.11 AirManager in the Boeing 757 aircraft (BAE Systems). Reproduced with permission from BAE Systems. See Plate 6 for the colour figure.

The system has been rigorously tested. All contaminants likely to be encountered in an aerospace application have been tested and independently assessed, typically:

- odours, perspiration, galley smells;
- biohazards – bacteria, moulds, spores, endospores and viruses;
- VOCs – chemicals potentially produced by the flying environment – Skydrol hydraulic fluid, anti-icing fluids, NOX/SOX, CO2, CO;
- organophosphates, TCP, unburned hydrocarbons, O1, CO.

Installation of the system can be achieved during overnight line maintenance. Replacement of the units will be carried out at 'C' Check intervals when the unit is exchanged for a new one and the old unit is sent for overhaul at a service centre.

3.6 Request for Further Information

Late in 2009 the European Aviation Safety Agency (EASA) published a discussion paper on cabin air quality and invited comments from anyone in the aviation industry with knowledge about the subject. It made available a series of questionnaires designed to collect evidence from pilots, cabin crew and operators as well as large aeroplane manufacturers and aviation authorities.

EASA is the central strand of the European Union's strategy for aviation safety and the aim of this ongoing project is to collect information which will lead to a clearer understanding of fume events and toxic air experiences involving large commercial aircraft.

If, on analysis of the data, EASA considers there to be a safety threat, it may create new airworthiness standards in order to limit the occurrence of future 'fume events'. The questionnaires, which are confidential, are the first step in a consultation process which could lead to these new recommendations, but EASA is also monitoring other concurrent projects, such as Cranfield University's study for the Department of Transport in the UK.

References

Australian Department of Defence (2005) Organophosphate and amine contamination of cockpit air in the Hawk, F-111 and Hercules C-130 Aircraft, DSTO publications, October.

European Standard EN 1822-1. Draft. High Efficiency Air Filters (EPA, HEPA, ULPA).

Lawson, C. (2010) *Wiley International Encyclopaedia of Aerospace*, Chapter eae 468, John Wiley & Sons, Ltd.

Mackenzie-Ross, S.J. (2008) Cognitive function following exposure to contaminated air on commercial aircraft. A case series of 27 airline pilots seen for clinical purposes. *Journal of Nutritional and Environmental Medicine*, **17** (2), 111–126.

Mackenzie-Ross, S.J., Harper, A.C. and Burdon, J. (2006) Ill health following reported exposure to contaminated air on commercial aircraft: psychosomatic disorder or neurological injury? *Journal of Occupational Health & Safety: Australia & New Zealand*, **22** (6), 521–528.

Marais, K. and Waitz, I.A. (2009) Air transport and the environment, in *The Global Airline Industry* (ed. P. Belobaba *et al.*), John Wiley & Sons, Ltd, pp. 405–436.

Moir, I. and Seabridge, A.G. (2008) *Aircraft Systems*, 3rd edn, John Wiley & Sons, Ltd.

Starmer-Smith, C. (2009a) 'Toxic' cabin air found in new plane study. *Telegraph Travel*, 14 February.

Starmer-Smith, C. (2009b) Illness among cabin crew heightens toxic air fears. *Telegraph Travel*, 18 July.

Yeoell, L. and Kneebone, R. (2003) On-Board Oxygen Generation Systems (OBOGS) for In-service Military Aircraft – The Benefits and Challenges of Retro-Fitting, Internal Honeywell report

Further Reading

AAIB Bulletin No. 7/2005: EW/G2004/11/08 & EW/G2004/11/12: Boeing 757-236, G-BPEE, 12, 16 and 23 November 2004.

Abou-Donia, Mohamed B. (2009) Chemical-Induced Brain Injury. Presentation given to GCAQE, London, http://www.gcaqe.org (presentations), (accessed February 2010).

Airline in-flight magazines, e.g. *Holland Herald* (KLM) , *High Life* (British Airways).

BAe Systems verbal evidence to Australian Senate Inquiry 2000.

CASA: Air Safety & Cabin Air Quality – Jim Coyne – A/g General Manager Manufacturing, Certification & New Technologies Office, 2007 presentation.

Commonwealth of Australia Senate Hansard. Monday, 13 August 2007 and Thursday, 20 September 2007, Aircraft Cabin Air Quality – Senator O'Brien.

Contaminated Air Protection Conference Proceedings (2005) Imperial College, London, 20–21 April 2005.

Cummin, A.R.C. and Nicholson, A.N. (2002) The cabin environment, in *Aviation Medicine and the Airline Passenger* (eds A. Cummin and A. Nicholson), Arnold.

Environmental Protection Agency (2006) EPA Green Book, www.epa.gov/air (accessed February 2010).

FAA (2002) faa.gov/safety/programs_initiatives/aircraft_aviation/cabin_safety (accessed February 2010).

German Ministry of Transport, Secretary of State Ulrich Kasparick, Question to MP Winfried Hermann of Bundnis90/Green Party in regard to contaminated cabin air on board civil airliners, printed matter 16/12023, 3 March 2009.

Gulf War Illness and the Health of Gulf War Veterans – Scientific Findings and Recommendations – Research Advisory Committee on Gulf War Veterans Illnesses, US Department of Veterans Affairs, Washington, DC, 2008.

Hocking, M.B. and Hocking, D. (2005) *Air Quality in Airline Cabins and Related Enclosed Spaces*, Springer.

Michaelis, S. (2007) *Aviation Contaminated Air Reference Manual*, Susan Michaelis.

Michaelis, S. (2007) Letter from Captain Susan Michaelis to the 2007 UK House of Lords Inquiry, available at: www.publications.parliament.uk (accessed February 2010).

Michaelis, S., Winder, C., Hooper, M. and Harper, A. (2008) Critique of the UK Committee on Toxicity Report on Exposure to Oil Contaminated Air on Commercial Aircraft and Pilot Ill Health, available at: www.aopis.org/ScientificReports.html (accessed February 2010).

Mobil Oil Corporation (1983) *Mobil Jet Oil II*, Environmental Affairs and Toxicology Department, New York, Correspondence, available at www.exxonmobil.com (accessed February 2010).

Nicholas, J.S., Lackland, D.T., Dosemechi, M. *et al.* (1998) Mortality among US commercial pilots and navigators. *Journal of Occupational & Environmental Medicine*, **40** (11), 980–985.

NTP Chemical Repository data, Radian Corporation, 29 August 1991, Tricresyl phosphate.

Quick Reference Guide for Health Care Providers. Health impact of exposure to contaminated supply air on commercial aircraft. San Francisco Division of Occupational Health and Environmental Medicine.

Rayman, R.B. and McNaughton, G.B. (1983) Smoke/fumes in the cockpit. *Aviation, Space and Environmental Medicine*, **54**, 738–740.

Rolls Royce, Germany (2003) *BRE Air Quality Conference*, London.

SAE Aviation Information Report: 1539, 30 January 1981.

Senate Rural & Regional Affairs & Transport References Committee (2000) Air Safety & Cabin Air Quality in the BAe 146 Aircraft. Final report, Parliament of Australia, Canberra, October 2000, Sections 5.31–5.32.

Turner v. *Eastwest Airlines* [2009] NSWDDT 5 May 2009, Australian Court.

UK Government Hansard: 66599, 4 February 1999, column 737.

UK COT Report: Long term sequelae of acute poisoning (1999) Committee on Toxicity of Chemicals in Food, Consumer Products and the Environment: Organophosphates: Executive Summary, Department of Health, London.

UK HSE: Organophosphates: HSE: MS17: Medical aspects of occupational exposures to organophosphates. Draft revision 23, November 1998.

Useful Web Sites

aerotoxic.org
easa.europa.eu
flightglobal.com
hepafilter-pro.com
ohrca.org/healthguide

4

Deep Vein Thrombosis

4.1 Introduction

Deep vein thrombosis (DVT) is a well-described symptom of long-haul commercial flying, and one of the few serious health problems that can affect occasional air travellers as well as frequent flyers.

DVT is where a clot forms within a deep vein – usually in the lower leg. Symptoms are tenderness and swelling of the affected part. DVT is often associated with long periods of immobility and can be detected through medical testing. It can be treated, but it can become much more serious when associated with thrombo-embolism, when the clot from DVT breaks off and travels to the lungs, where it becomes stuck and blocks the flow of blood. This is known as pulmonary embolism, and symptoms include chest pain and breathing difficulties. If not treated promptly, it can prove fatal.

The deep veins of the lower leg, illustrated in Figure 4.1, are a common site for DVT. Immobility for long periods in cramped cabin conditions can restrict blood flow and may increase the risk.

DVT – or venous thrombo-embolism to give its full name – after air travel was first recorded in 1954, but the problem of immobility causing blood clots in major blood vessels was recognised as far back as the First World War, where there were descriptions of soldiers suffering venous thrombosis after being cramped in trenches for long periods. In 1940 Dr Keith Simpson investigated the mysterious deaths of 23 people who had slept in deckchairs in the London Underground during air raid warnings

Air Travel and Health: A Systems Perspective   Allan Seabridge and Shirley Morgan
© 2010 John Wiley & Sons, Ltd

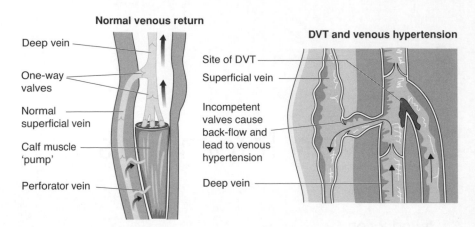

Figure 4.1 The deep veins of the lower leg (Blamb). Reproduced with permission from Shutterstock Images. See Plate 7 for the colour figure.

the year before. Simpson established that they died from blood clots to the lungs, caused by prolonged immobilisation in a seated or at best semi-reclined position, and the authorities quickly replaced the deckchairs with bunk beds, which allowed people to lie down and stretch out. The deaths stopped.

In 1968 Drs P.H. Beighton and P.R. Richards wrote a paper alerting the airlines to the dangers of DVT from long-haul flights (Beighton and Richards, 1968). In 1985 Yvonne Hart, D.J. Holdstock and William Lynn from Ashford Hospital, near Heathrow, wrote to *The Lancet*: 'We see a steady stream of illnesses which have developed in flight. The major manifestation of the illness may not occur until after disembarkation. We have seen several patients with thrombo-embolism presenting in this way, with a near-fatal outcome in one case' (Scott *et al.*, 1985).

Both commercial and military pilots spend long periods of time confined to their seats with little opportunity to 'stretch their legs'. Passengers can also be forced to spend many hours in their seats, and concerns about in-flight security and safety mean passengers are not encouraged to move around the aircraft cabin. This chapter will examine the evidence and anecdotes and discuss methods of reducing the occurrence of a condition that has been dubbed by the popular press 'economy-class syndrome' – actually a misnomer, as people who fly first class have been killed by it. In fact a Japanese study by Dr Noritake Hata found 70% of travellers in economy class, 25% in business class, 5% in first class, and one pilot who suffered from clots. The ratio of victims was the same in all sections of the aircraft, including the flight deck, and a 2003 study by Gianni Belcaro found 4.5% of frequent business-class travellers developed one or more clots per year (Belcaro *et al.*, 2003, 2004).

4.1.1 How Common Is It?

Hundreds of reports and journal articles have been written about the problem of DVT following flight. Most authors agree that the major cause in traveller's thrombosis is a 'sludging' of the blood in the blood vessels due to sitting in a confined space, with limited mobility. Although other cabin factors have been investigated – particularly low pressure and reduced oxygen levels – most advice on preventing DVT focuses on avoiding restricted blood flow in the lower limbs.

Estimating just how many people have been affected by post-flight DVT is difficult. It can take up to 10 days before a blood clot breaks off and moves to the lungs, which means that death from pulmonary embolism can occur many days after the flight and may not be directly attributed to travel. Another complication in compiling accurate statistics is that DVT can happen for many other medical reasons, and everyone has their own personal set of 'risk factors', which can predispose them to DVT.

Research carried out by a consortium of medical scientists under the auspices of the World Health Organization (29 June 2007) has found one case of DVT for every 6000 journeys that lasted four hours or more. But in most cases it is still difficult to say whether the flight itself caused the DVT or whether it happened for other reasons.

The aim of the WRIGHT (WHO Research Into Global Hazards of Travel) project was to confirm that the risk of DVT and embolism is increased by air travel and to assess the level of risk, the impact of other factors and preventive measures. Several studies were performed during Phase I of the WRIGHT project. The findings of the epidemiological studies indicate that the risk approximately doubles after a flight of more than four hours. This risk increases with the duration of the journey and with multiple flights within a short period.

The risk of blood clots also increases significantly in the presence of other known factors such as obesity, extremes of height, use of oral contraceptives and the presence of blood abnormalities.

Following on from the WRIGHT report, in 2007 calls were made for more research to explore ways to prevent DVT. The Aviation Health Working Group (AHWG), chaired by the Department for Transport, with representatives from the Civil Aviation Authority, Health and Safety Executive and the Department of Health, wants to ensure that further research ties in with that directed by the World Health Organization (WHO).

4.1.2 How Long Is a Long Flight?

As we have seen, there is evidence to support the view that the longer the trip, the greater the risk of thrombosis, but it is not possible to say how short a flight must be for it to be 'safe'. Remember that it is the immobility, not the mode of transport, that adds risk. Prolonged immobility in a train, car or bus could potentially cause DVT, as could sitting in a cramped theatre, but it is obviously easier to break journeys or to get up and walk around in such cases.

A study by the UK-based Aviation Health Institute found that 17% of flight-related DVT cases occurred in association with short flights. It has also been demonstrated that the duration of travel is not linked to the severity of the thrombosis suffered (Parsi *et al.*, 2001).

The *Independent* newspaper (17 October 2003) published interim findings on the incidence of DVT in high-risk passengers as a result of a short-haul flight (London–Rome) which is of less than three hours' duration. The authors found that 4.3% of 568 passengers developed clots, which were detected by an ultrasound scan. Two of the passengers in the sample went on to suffer a pulmonary embolism. The lead researcher, Professor Gianni Belcaro, of G. d'Annunzio University in Italy, said the studies suggested that most blood clots develop in the first two to three hours of a journey and grow larger and more dangerous with time.

4.2 The Environment

The commercial aircraft environment is the inhabited portion of the aircraft reserved for the flight crew, the flight deck or cockpit, and that reserved for the passengers and the cabin crew, the passenger cabin. The flight deck is usually sealed off from the passenger cabin for reasons of security, with access granted to the cabin crew only on visual identification. The pilot and first officer are confined to the flight deck for the duration of the flight except for visits to the toilet and to change with the reserve pilot on long-haul flights.

The passenger cabin is provided with seats in various sections in line with fare classes, variously first, business, premium economy and economy, although the designations change with each airline's preferred terminology and marketing strategy. The cabin is equipped with toilets and one or more galleys where food is prepared. Food and drink are served to passengers at their seats from trolleys by the cabin crew.

The cabin atmosphere is controlled to maintain a pressure 'altitude' of between 5000 and 10 000 ft and the cabin air cleanliness, temperature and humidity levels are controlled. If water extraction is too severe the cabin conditions can lead to dehydration, which may be a factor in DVT. As an example of extreme conditions the Nimrod MR2, a UK maritime patrol aircraft (MPA), had a reputation for evaporating cups of coffee far quicker than crews could drink them. For more information see Chapter 3 on air quality.

The large military aircraft designed for missions such as ground surveillance, airborne early warning and maritime patrol are usually based on a commercial airliner platform. The flight deck is similar, whilst the cabin will be equipped with workstations to allow the crew to do their tactical tasks. The cabin is equipped with toilets and galley and conditions are maintained much as the commercial types. For very long missions bunk beds are often provided to allow reserve crews to rest. The mission crew have freedom to move around, although most of their time will be spent at their workstations.

The small, fast-jet military environment is different. The cockpit is designed for one person, two in the case of trainers and some aircraft requiring a pilot and navigator. The cockpit is designed so that the crew members remain seated at all times, usually on an ejection seat. They are strapped in very tightly and there is no room for exercise.

4.3 Aircraft Environments

4.3.1 Commercial Aircraft

Commercial aircraft are often classified as single aisle or twin aisle as illustrated in Figure 4.2. Seats are aligned in rows on either side of the aisles at a pitch or seat interval, largely determined by the class in which passengers are flying. The number of seats on each side of the aisle varies according to the body diameter of the aircraft. Individual airlines have their own policy on the optimum seat pitch to meet their operating costs. Some airlines have a deliberate policy to attract customers by declaring a more comfortable seating arrangement than their competitors.

Movement in the aisles is restricted, since they are primarily a route for cabin attendants, trolleys for meal service and reaching the toilets and exits. Passengers are not encouraged to linger as they obstruct the aisles for other people. This severely restricts the use of aisles as a means of exercise.

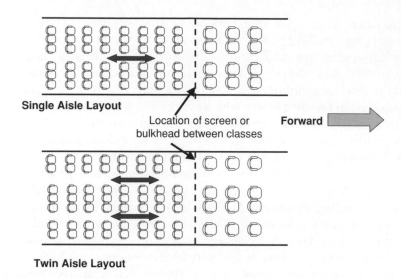

Figure 4.2   Illustration of single- and twin-aisle cabins.

Movement is further restricted by the need to wear a seat belt for most of the flight – 'remain seated for your own comfort and when the captain has turned on the seat belt signs'. This is to protect passengers from the effects of sudden turbulence. The aisle is difficult to negotiate, as illustrated in Figure 4.3.

The in-flight magazine will often encourage passengers to exercise whilst seated by stretching, extending and rotating the joints of feet, ankles, knees, shoulders, legs, back and arms. This is said to improve well-being or to keep 'feeling great both during and after flight'.

The range of seats varies considerably according to airline preference, class of cabin, type of aircraft service – regional, short haul, long haul. Some examples are shown in Figure 4.4. On long-haul flights the ability to recline and lift the legs is essential for a comfortable night's sleep. The use of flat beds in first class allows the occupant to lie flat, stretch out fully and lie on one side – many people's preferred sleeping position, which is just about impossible in a reclining seat.

4.3.2 Large Military Aircraft

Large military aircraft play a number of roles including troop transport, in-flight refuelling, ground surveillance, airborne early warning and maritime patrol. Many of these types are based on commercial airliners which have

B737 cabin – short haul
Allan Seabridge

Single aisle layout – Paul Prescott

Twin aisle layout – Oksana Perkins

Figure 4.3 Examples of an occupied economy cabin in flight. Reproduced with permission from 1. Allan Seabridge 2&3. Courtesy of Shutterstock Images. See Plate 8 for the colour figure.

been modified for a specific use. Common platforms in use today are the Boeing 707, Comet, Lockheed Orion and Global Express.

These aircraft carry a crew of 12 to 20 people whose task is to perform the mission for which the aircraft is designed. For surveillance roles this entails sitting at a workstation throughout the mission, monitoring pictorial and graphical data on large screens and making decisions based on that information. For this the crew are provided with airline standard seats, mounted on rails, which allows some lateral and rotational movement, so that they can be aligned fore and aft for take-off and landing. Full harness seat belts are usually provided. There is some mobility in this 'office' environment. Crew members may wish to speak to other crew members, go to the galley or toilet and refer to manuals. Food and drink are provided, but alcoholic beverages are not available. Figure 4.5 shows the mission crew in the Nimrod MRA4; the crew sit in line looking outboard to the left of the aircraft.

Standard class seats - Magicinfoto

Business class seats –
Vladimir Sazonov

Business class seats – Petronito G. Dangoy Jr

Figure 4.4   Examples of seats available in economy, business and first class. Reproduced with permission from Shutterstock Images. See Plate 9 for the colour figure.

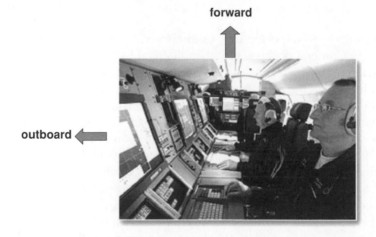

Figure 4.5   The mission crew in a Nimrod MRA4 (BAE Systems). Reproduced with permission from BAE Systems.

Materiel transport types, which often carry troops as well, are mostly built for that specific purpose and include Lockheed C-130, Boeing C-17, Antonov AN-14. The seats for troops are rudimentary canvas seats for what is expected to be a single and occasional journey.

4.3.3 Fast-Jet Military Aircraft

The military fast-jet pilot does not enjoy a vast office – it can be said that the pilot wears the aircraft rather than enters it. The cockpit volume is constrained by the front panels, side consoles and rear bulkhead. The pilot sits on an ejection seat which is installed with a seat back angle that is optimum for the view of the front panels, tolerance to high-g manoeuvres and safe escape. The seat back angle is fixed as is the seat position, and the pilot wears a full harness restraint, tightly secured to ensure safe ejection. The preferred position is with the left hand on the throttles and the right hand on the control column or stick, where the majority of instinctive controls are located. This technique is known as hands on throttle and stick (HOTAS).

The cockpit of the Eurofighter Typhoon showing the pilot seated in an ejection seat is shown in Figure 4.6.

4.4 The System

4.4.1 Commercial Aircraft

If cramped conditions are a factor in causing DVT, then seat pitch must play a part. Passenger seat pitch (seat pitch is the horizontal distance between similar points on two seats situated one behind the other) varies in accordance with the class of ticket, although some airlines offer wider seat pitches as part of their marketing strategy. It is generally only first and business classes that offer full-length beds on long-haul routes, otherwise the seat angle is fairly restricted. In economy class the opportunity for movement without disturbing fellow passengers is limited, especially in the centre rows that may have four seats. Trolleys used for passenger service, and the current terrorist threat, further discourage passengers from walking in the aisles.

British Airways attracted publicity when it decreased the distance between seats from 91 to 79 cm. Seats on Qantas flights are about 81 cm apart. For Air Canada, the seats are spaced about 83 cm apart. The distance between seats is not identical even on the same plane. Exit and bulkhead seats have much more legroom, making them safer for potential traveller's thrombosis victims, but may have restricted recline angles. Comparative seat pitch information

Geoffrey Lee, Planefocus

BAE Systems

BAE Systems

Figure 4.6 View of the Eurofighter Typhoon cockpit (BAE Systems). Reproduced with permission from BAE Systems, Geoffrey Lee Planefocus.

for major airlines is available on the internet on various web sites. For a general comparison of airlines see www.aviation-health.org. For more detailed information related to different aircraft types try www.simplyquick.com.

The House of Lords Select Committee on Science and Technology (2000) recommended that the CAA implement the recommendations of its own research into aircraft seating standards, and increase the regulatory minimum distance between seats to at least 28.2 inches (71.6 cm). It also asked the government to review urgently the level of air passenger duty levied on 'premium economy' seating and explore ways in which the airlines can be encouraged to offer extra space to passengers for a modest premium.

4.4.2 Military Aircraft

In the restricted cockpit environment of a fast jet there is little or no opportunity for movement, and the professional pilot or crew member can expect to do many hours of flying each year. The sitting position is almost fixed with the pilot sitting in an ejection seat with hands on throttle and stick. This cramped position leaves no room for movement, certainly no room for exercise. Fast-jet missions are relatively short, an hour or two, but may be extended if the aircraft is on combat air patrol (CAP) or has its endurance extended by the use of in-flight refuelling. Two-seat trainer aircraft also have short missions, but the instructor may fly several training flights a day with different student pilots.

Military crew cannot exercise *in situ*, and therefore must maintain fitness constantly.

4.5 Health Issues

Symington and Stack, who reported an apparent increase in DVT in air travellers compared with non-travellers, first described 'Economy Class Syndrome' or 'Traveller's Thrombosis' in 1977. It was attributed to cramped seating in economy-class cabins and they realised that long periods of in-activity caused by lack of legroom could slow the circulation and produce oedema (leg swelling). In addition, bent knees compress the deep vein behind the knee, creating a potential 'hot spot' for clot formation. They also considered that low oxygen, low humidity and low cabin pressure could have a dehydrating effect that could make the blood flow more sluggish. This effect is worsened when air passengers drink alcohol or become dehydrated, and individual 'risk factors' such as smoking and being overweight further increase the risk of DVT.

Because of the difficulties in collating data, estimates of the number of deaths from post-flight DVT vary widely. A conservative estimate is considered to be about 100 a year in the UK, though David Derbyshire, medical writer for the *Daily Telegraph* (11 January 2006), says that doctors who carried out a study at Ashford Hospital in Surrey believe more than 2000 people a year die from the condition in the UK.

One hospital study in England found that at least 30 people died in a three-year period in the late 1990s of massive blood clots after arriving at Heathrow Airport on long-haul flights. An Australian surgeon has been quoted as saying 400 people a year landing at Sydney Airport suffer blood clots (CBS News, 2006).

The condition is not always fatal. A story in the *New York Times* in October 2000 estimated that 5 million Americans a year experience a blood-clotting thrombosis caused by prolonged immobility, resulting in 800 000 hospitalisations a year.

Looking at air travel in isolation, studies have shown that 3 to 5% of travellers develop clots in veins. Some, of course, are so mild as to cause no symptoms and the travellers are totally unaware that they have even had a clot. In 2001 *The Lancet* (8 September 2001, p. 838) published an analysis estimating that 1 million cases of DVT related to air travel occur in the US every year and that 100 000 of these cases result in death.

4.5.1 Cabin Altitude and Pressure

There are suggestions that aircraft cabin altitude and cabin pressure could be key risk factors in blood clotting. Could altitude, not immobility, be the primary cause of flight-induced DVT?

A Norwegian study published in *The Lancet* (Bendz *et al.*, 2000) put 20 young men in a hypobaric chamber, which simulated aircraft cabin altitude. Cabin pressures simulated an altitude of 5000 to 8000 ft in various aircraft types, reducing oxygen pressure from 98 to 79 mmHg, as calculated for a Boeing 747. It has been calculated that this can lead to 90% saturation of haemoglobin with oxygen, a figure that may be reduced even further by sleep and the effects of cramped conditions on respiratory mobility. Factors such as cabin humidity were also considered. It was found in the study that a substantial hour-by-hour increase in blood-clotting factors occurred in all of the healthy subjects. There was a two- to eight-fold increase in clotting factors, implying that all flyers are subject to this increased risk, and suggesting that those who succumb to DVT have a variety of risk factors deriving from their own genetic and physiological make-up and their environmental circumstances.

This enhanced chance of coagulation has also been demonstrated (Schobersberger *et al.*, 2002) in a study measuring coagulation factors on an actual long-haul flight. The effects were observed in all test subjects. The report concluded:

> Long haul flights induce a certain activation of the coagulation system. This activated coagulation could be a risk factor for venous thrombo-embolism (VTE) during long-haul flights mainly when other risk factors are present.

Dr John Marx at the Ashley Clinic in Melbourne has measured venous failure in the lower legs caused by cabin altitude. Studies show that a compression stocking rated at 20 mmHg (20 millimetres of mercury) doubles the venous return of blood from the leg to the heart. Dr Marx wondered if reducing pressure had the opposite effect, reducing venous return, and, if so, would the effect on air travellers be harmful? In the first eight minutes of a jetliner's climb, cabin pressure drops from about 760 to about 560 mmHg, a difference of 200 mmHg – 10 times greater than that of a 20 mmHg compression stocking and in the opposite direction.

Using a technique for evaluating vein function, Marx (2009) took measurements while ascending a mountain, with air pressure at the top about equal to cabin altitude. At an altitude of 1380 metres (4500 ft), venous blood flow had fallen by about 50%. At the top of the mountain (2228 m (7400 ft)), venous blood flow had been further reduced to only 30% of normal. 'Our subjects were tipped into severe venous failure', he says, suggesting that all air travellers will be prone to stagnant blood during periods of reduced barometric pressure.

An interesting study led by Anja Schreijer at Academic Medical Centre and Leiden University Medical Center (the Netherlands) compared thrombin levels in air travellers with those of people who simply sat on the ground and watched movies for eight hours (Schreijer et al., 2008). Thrombin is part of the body's clotting mechanism and the researchers found a 223% rise in levels due to travelling compared with 46% rise due to immobility, suggesting that a mechanism other than immobility caused air travellers to be at an increased thrombotic risk.

These 'other factors' at work could suggest why 'economy-class syndrome' is such a misnomer. Pressure at the front of the aircraft cabin is the same at the rear, and everyone is flying at the same altitude, making business- and first-class passengers just as likely to develop DVT as their economy-class companions.

4.5.2 So Who Is at Risk?

The risk of developing DVT (not only after air travel) is very strongly linked to age. It is rare (though not unheard of) in young people and common in the elderly. Women taking oral contraceptives are also much more vulnerable to DVT and those on oestrogen replacement therapy also have a higher risk – partly because they are older. Women who have recently given birth are higher risk and some people believe that pregnant women run such an increased risk that they should not fly at all, since preventative anti-coagulant therapy in the event of DVT could have serious consequences for the foetus.

The consequences of DVT can be particularly worrying for some other groups – diabetics, for example – as the effective treatments may be unsuitable, given their underlying conditions.

Pilots are at risk, too, and the first sign of a problem can be fainting. Dr David McKenas, medical director at American Airlines, says that the most common causes of sudden pilot incapacitation are cardiac arrest, arrhythmia and fainting (all of which can be caused by a blood clot in the lung). With DVT, pilots, just like other victims, can soldier on with symptoms for a considerable time. During those days and weeks they are at risk of sudden collapse, and are a danger to themselves and their passengers. Pilots are also at risk of sudden collapse due to arterial clots causing heart failure or stroke. Having had DVT does not disqualify a pilot, but everyone in the industry would say it is better to avoid getting one.

Pilots are increasingly concerned about thrombosis and some are visiting internet sites to express their alarm. One said: 'I can confirm that aircrew have in the past, and still do, suffer from DVT', the pilot wrote. 'I suffered a pulmonary embolism at the age of 37.'

The CAA confirmed that a number of pilots had suffered from DVT in the past. 'Pilots have very strict medicals and some of them have had blood clots', a spokesman said. 'However, we do not know if it was anything to do with flying.' (See, for instance, www.guardian.co.uk, 14 June 2001.)

Being young and healthy with apparently no underlying health problems or predispositions is no guarantee of safety. DVT first hit the front pages in September 2000 after the death of Emma Christoffersen, a 28-year-old Briton who died after returning to London from Australia, where she had attended the Olympic Games. Christoffersen collapsed at Heathrow Airport following the 20-hour trip and died before she reached hospital. It was found that she had developed a clot in her leg on the Qantas flight; the clot dislodged and made its way to her heart.

4.5.3 Summary of Risks

Despite the horror stories, generally the risk of developing serious DVT when travelling is very small unless you have at least one of the other risk factors.

Lower risk factors:

- Hormone therapy or use of the contraceptive pill
- Obesity (a body mass index (BMI) of more than 30)
- Varicose veins
- Inflammatory bowel disease
- Age 40+
- Recent leg trauma.

Higher risk factors:

- Family history of venous thrombosis
- Pregnancy or post-partum
- Age 60+
- Current or historic malignancy
- Serious lung problems
- Recent surgery or limb immobilisation (e.g. plaster cast)
- Recent stroke
- Known blood-clotting problems.

Extremely high risk factors:

- Personal history of venous thrombosis

4.6 System Implications

4.6.1 Litigation

DVT is a topical subject in the world of commercial aviation with passengers considering litigation against airlines. The number of court cases resulting from incidences of DVT is growing. Some travellers have cited the Warsaw Convention of 1929, which holds airlines liable for damages when passengers are injured by an accident, as grounds to file negligence claims. The ongoing debate is whether a blood clot is a preventable event, or an individual reaction to normal flight operations.

Families of the victims argue that the convention should be seen as a form of consumer protection to allow passengers to seek redress if damaged by acts or omissions by the carriers. The airlines maintain that it should only apply to damage caused by mechanical faults. Court rulings have been inconsistent internationally.

Early cases presented in US district courts began in Texas (*Reynolds* v. *American Airlines*, 2002) and California (*Miller and Wylie* v. *Continental Airlines*, 2003). Since then, hundreds of cases have been filed domestically and internationally.

In the UK, such cases have attracted wide publicity, as in the case of former pub landlady Val Clark, who flew to the US for a holiday with friends. Cramped conditions on the return flight led to blood clots, a lengthy spell in hospital and eventually the loss of her leg. She successfully sued North West Airlines and accepted an out-of-court settlement following her ordeal.

In *Blansett* v. *Continental Airlines* in 2002, Judge Kent ruled that Shawn Blansett's stroke suffered aboard a Continental flight to London was an accident as defined by the Warsaw Convention.

The Times in 1998 reported that Judge Gareth Edwards, QC, said that airline seats should occupy a space of at least 34 inches (83.3 cm) when he upheld a compensation award made to a company director who had suffered 'intolerable' discomfort on a flight from Canada, when confined to a 29 inch (73.67 cm) space. In many cases, inconclusive medical evidence has meant that airlines have refused to accept responsibility, and the uncertainty around post-flight DVT and who is responsible underlines the importance of ongoing medical studies into the cause-and-effect relationship between air travel and thrombosis.

4.6.2 Preventative Measures

Several airlines, British Airways included, have taken part in studies of DVT.

There are dozens of web sites, travel magazines and airline in-flight journals which have advice on how to protect against DVT.

4.6.3 Advice to Passengers

The most common advice is to:

- Drink plenty of non-alcoholic drinks during the flight (but avoid caffeine).
- Try not to sit with legs crossed.
- Walk around the cabin when you can.
- Try to stay awake and do not take sleeping pills.
- Stand up and stretch your arms and legs, flex and rotate the ankles and feet regularly while sitting down.
- Wear loose-fitting clothes that do not restrict blood flow.

In addition:

- If you have long legs or are either very tall or very short (taller than 1.9 m or shorter than 1.6 m) find an airline with the largest seat pitch possible, or request a seat by the emergency doors or bulkhead where there is more legroom.
- Smokers increase their 'personal' DVT risk factor – a long flight might be a good incentive to quit.

- Buy a pair of elastic 'compression' socks, which should be worn for the duration of the flight.
- If you can tolerate aspirin, then half an adult aspirin tablet can help to reduce the risk of blood clots.

There are many variations on the same themes – but some of these 'tips' carry a note of caution. Staying well hydrated and avoiding alcohol certainly sounds like sensible advice, but a study in Japan by Hamada *et al.* (2002), published in the *Journal of the American Medical Association*, found that subjects who drank one cup of water per hour during a nine-hour flight experienced *increased* blood viscosity. Interestingly, the study found that those who drank an electrolyte fluid (similar to a sports drink) had no increase in blood viscosity and no increase in urinary output.

Alcohol is proved to have a dehydrating effect during flight – but what about red wine? Moderate consumption is considered to be good for the health of blood vessels because it reduces the 'stickiness' of blood platelets. A Polish research group found that the resveratrol present in the human diet (red wine has significant amounts) may be an important compound responsible for the reduction of platelet adhesion (Olas *et al.*, 2002). Could a glass of red help prevent flight-related DVT?

Though the use of aspirin is often recommended, care should be taken with any drugs used before travelling. Prescription blood 'thinners' such as Warfarin should not be taken, except on medical advice.

The jury is still out on the benefits of compression flight socks. They work by applying gentle pressure to the ankle, helping blood to flow around the body by squeezing it up towards the heart. They have certainly grown in popularity in recent years and anyone undergoing abdominal surgery or facing a prolonged stay in hospital may be routinely issued with a similar pair. It has been reported that at least two international soccer teams wear compression stockings on long-haul flights, and several studies have shown that wearers of compression stockings are dramatically less susceptible to DVT than those not wearing them. One study has shown a complete elimination of the risk in compression stocking wearers (*The Lancet*, 12 May 2001, pp. 1485–1488) but compression stockings did not help four of nine people who developed DVT and were part of a study by Hughes *et al.* (2004) involving 878 long-distance air travellers. They found five cases of DVT and four pulmonary embolisms.

The evidence is inconclusive, and a rushed purchase in the duty-free shop may be ill advised. It is vital that the stockings are the correct size and are worn properly so that they do not 'bunch' in one place, causing a too-tight band around the leg. People with serious varicose veins should take medical advice before wearing compression stockings.

If compression socks do work, then the merits of issuing them routinely to flight crew would be obvious. However, one pilot, speaking to Airhealth.com, said he would be reluctant to wear them because it might raise questions about his overall fitness.

4.6.4 Continuing Work

Because there are mixed messages when it comes to DVT and passenger health, there are calls for more consistent advice for passengers on the risks associated with self-medicating or using other techniques with the intention of preventing DVT.

The Select Committee on Science and Technology recognised that there is a lack of data on this topic and recommended that an epidemiological research programme should be commissioned by the Department of Health. This recommendation was made in respect of commercial air travel, but there is likely to be some correlation to the military pilot situation (House of Lords Select Committee on Science and Technology, 2000). The select committee has also recommended an end to the phrase 'economy-class syndrome' on the basis that it is not accurate.

The report from the Air Transport Medicine Committee (Traveller's Thrombosis) concluded that any association between symptomatic DVT and air travel is weak based on current evidence (Bagshaw, 2001). The committee supports the select committee recommendation to support a research programme. This report also mentions a theory which suggests that hypoxia may be a factor in DVT, although current research does not support this view.

Legroom on commercial flights will always be an issue, and with airlines coming under increasing financial pressures, the temptation must be to fly with as many seats as is possible. But could cinema-style 'flip seats' give passengers more space and reduce the risk of DVT?

Prototypes were unveiled in Hamburg and are designed to make standing up much easier, according to the German manufacturers Aida, giving passengers more room. The company claims that flip seats would speed up boarding and disembarking and make it easier to stand and stretch the legs mid-flight, to reduce the risk of DVT. Thomsonfly is reported to have expressed an interest.

Airlines are seeing the benefits of improving seating conditions and a recent announcement by Air New Zealand declares that economy-class passengers are to be offered beds but will have to buy three seats to qualify. Air New

Zealand said its 'Skycouch' will use three economy seats that unfold to create a space where children can play or people can sleep. The deal is aimed at couples and families and the price will equate to about two and a half seats.

To create the bed, leg rests rise to fill the space between seats. A thin mattress is put on top and pillows are provided. The result is a bed across three seats, although it is not completely flat due to the seat contours.

The airline will start installing the 'Skycouch' in 2010 when it takes delivery of the first of its new Boeing 777-300 aircraft (www.nzherald.co.nz, January 2010).

References

Bagshaw, M. (2001) Traveller's Thrombosis – A Review of Deep Vein Thrombosis associated with air travel. Air Transport Medicine Committee. *Aerospace Medical Association*, **72**, 848–851.

Beighton, P.H. and Richards, P.R. (1968) Cardiovascular disease in air travellers. *British Heart Journal*, **30**, 367–372.

Belcaro, G., Cesarone, M.R., Nicolaides, A.N. *et al.* (2003) The LONFLIT4-VENORUTON study: a randomized trial prophylaxis of flight-edema in normal subjects. *Clinical and Applied Thrombosis/Hemostasis*, **9** (1), 19–23.

Belcaro, G., Cesarone, M.R., Rohdewald, P. *et al.* (2004) Prevention of venous thrombosis and thrombophlebitis in long-haul flights with Pycnogenol. *Clinical and Applied Thrombosis/Hemostasis*, **10** (4), 373–377.

Bendz, B., Rostrup, M., Sevre, K. *et al.* (2000) Association between acute hypobaric hypoxia and activation of coagulation in human beings. *The Lancet*, **356**, 1657–1658.

CBC News Online (2006) Indepth: Health – The perils of traveller's thrombosis. 11 January.

Hamada, K., Doi, T. and Sakurai, M. *et al.* (2002) Effects of hydration on fluid balance and lower-extremity blood viscosity during long airplane flights. *Journal of the American Medical Association*, **287**, 844–845.

House of Lords Select Committee on Science and Technology, 5th Report (15 November 2000) Air Travel and Health, Chapter 1: Summary and Recommendations.

Hughes, R., Weatherall, M., Wilsher, M. and Beasley, R. (2004) Venous thromboembolism in long-distance air travellers. *The Lancet*, **363**, 896–897.

Marx, J. (2009) Altitude Induced Venous Failure. Bupa factsheet.

McKenas, D. (American Airlines) www.airhealth.com (accessed February 2010).

Olas, B., Wachowicz, B., Saluk-Juszczuk, J. and Zieluski, T. (2002) Effect of reseveratol, a natural polyphenolic compound, on platelet activation induced by endotoxin or thrombin. *Thrombosis Research*, **7** (3–4), 141–145.

Parsi, K., McGrath, M.A. and Lord, R.S. (2001) Traveller's venous thromboembolism. *Cardiovascular Surgery*, **9** (2), 157–158.

Air Travel and Health

Scott, R., Hart, Y., Holdstock, D.J. and Lynn, W. (1985) Medical emergencies in the air. *The Lancet*, **325** (8424), 353–354.

Schoberberger, W., Fries, D., Mittermayr, M. *et al.* (2002) Changes of biochemical markers and functional tests for clot formation during long-haul flights. *Thrombosis Research*, **108** (1), 19–24.

Schreijer, A., Cannegieter, S.C., Doggen, C.J.M. and Rosendaal, F.R. (2008) The effect of flight-related behaviour on the risk of venous thrombosis after air travel. *British Journal of Haematology*, **144** (3), 425–429.

Simpson K. (1940) Shelter deaths from pulmonary embolism. *The Lancet*, **ii**, 744.

Symington, I.S. and Stack, B.H. (1977) Pulmonary embolism after travel. *British Journal of Diseases of the Chest*, **71** (2) 138–140.

Further Reading

Department of Health (March 2007).

Kelman, C.W., Kortt, M.A., Becker, N.G. *et al.* (2003) Deep vein thrombosis and air travel: record linkage study. *British Medical Journal*, **327**, 1027.

Kos, C.A. National Alliance for Thrombosis and Thromboplilia. Stopetheclot.com (accessed February 2010).

Mckeown, M. (2003) Research Report on Deep Vein Thrombosis in Air Travellers. *International Health News*, No. 142.

Mittermayr, M., Fries, D., Gruber, H. *et al.* (2007) Leg edema formation and venous blood low velocity during a simulated long-haul flight. *Thrombosis Research* **120** (4), 497–504).

New York Times, October 2000.

Observer, Sunday 14 January 2001.

Schoberberger, W., Mittermayr, M., Fries, D. *et al.* (2007) Changes in blood coagulation of arm and leg veins during a simulated long-haul flight. *Thrombosis Research*, **119** (3), 293–300.

Scurr, J.H., Machin, S.J., Bailey-King, S. *et al.* (2001) Frequency and prevention of symptomless deep-vein thrombosis in long-haul flights: a randomised trial. *The Lancet*, **357**, 1485–1488.

Useful Web Sites

airhealth.org
www.nzherald.co.nz

5

Noise and Vibration

On the face of it noise should not be a major problem on most commercial aircraft – modern technology engines are quiet and aircraft sound insulation is very good. It was not always like this – the days of propellers driven by air-aspirated engines and turbo-props before the turbo-fan engine came along did introduce high noise levels and vibration into the cabin environment.

The military fared no better back then, the use of the Shackleton in RAF service as a maritime patrol aircraft providing some instances of hearing impairment. This aircraft was essentially an airframe designed for the Second World War and adapted for peacetime use in a role with a long mission duration. An example of the stoical nature of RAF aircrew in this situation is exemplified by Harry Jones: 'We all had ringing in the ears after each trip, but after a good few beers and good night's rest we were OK.' At age 72 Harry says he 'has had tinnitus and low tolerance of noise for over two years, which I put down to eight years on multi-engined aircraft with very high noise and vibration levels, but the doctors put down to age'. He wryly observes that the beer does not work any more.

The major source of noise and vibration on the Shackleton was the engine, in this case four Gryphon engines with contra-rotating propellers, as shown in Figure 5.1, coupled with the fact that the crew only had earphones.

Another observation comes from the Editor of *Growler* (the Shackleton Association magazine):

I suffer from tinnitus but it sounds exactly like the 115 V 400 Hz AC supply singing diode about a foot above the routine navigator's left ear on the

Air Travel and Health: A Systems Perspective   Allan Seabridge and Shirley Morgan
© 2010 John Wiley & Sons, Ltd

Figure 5.1 The Shackleton and its noise-generating engines (Allan Seabridge/ Manchester Museum of Science and Industry). Reproduced with permission from Allan Seabridge. See Plate 11 for the colour figure.

> Nimrod. I flew for 16 years on the Nimrod and only six years on the Shacks marks 2 and 3. Although John Bussey mentioned that he had an award for hearing loss due to the Shack (high tone deafness), I think that my hearing problems are just old age.

Doug Cox, a former Shackleton operator, also suffers from hearing loss after many years of tinnitus. One thing is clear: it is in later life after service retirement that hearing loss which may be attributed to aircraft noise becomes apparent.

A useful description of the subject is as follows. Tinnitus (from the Latin *tinnitus*, meaning 'ringing') is the perception of sound within the human ear in the absence of corresponding external sound. It is not a disease but a symptom from a range of underlying causes that can include ear infections, foreign objects or wax in the ear, nose allergies that prevent (or induce) fluid drain and cause wax build-up. Tinnitus can also be caused by natural hearing impairment (as in ageing) and as a side effect of some medications. However, the most common cause of tinnitus is noise-induced hearing loss.

Tinnitus is often defined as a subjective phenomenon and is difficult to measure using objective tests, such as by comparison with noise of known frequency and intensity, as in an audiometric test. The condition is often rated clinically on a simple scale from 'slight' to 'catastrophic' according to the practical difficulties it imposes, such as interference with sleep, quiet activities and normal daily activities.

Sources of noise and whole-body vibration from aerodynamic turbulent flow, engine intake, cabin cooling systems and headset noise, particularly from fast jets and vertical lift aircraft, can combine to produce noise levels in excess of those permitted in the workplace. The result is limitation of pilot flying hours or the appearance of chronic hearing problems.

5.1 The Environment

The aircraft environment subjects the whole body to vibration, particularly low-frequency vibration. Studies in occupational populations exposed to long-duration, high-amplitude, low-frequency noise vibration (110 dB, 100 Hz) postulate that a number of symptoms can arise, sufficient to identify a disease – namely, vibroacoustic disease (Alves-Pereira and Castelo Branco, 1999). The symptoms include disorder of joints and muscles, especially the spine, disorder of the circulation (hand–arm vibration), impairment of vision and/or balance. Some of these are similar to some symptoms experienced in aircraft use, although the research currently does not include pilots. Although there is some debate as to the validity of the claim that this is an occupational disease, the symptoms are interestingly similar to those experienced by aircrew but attributed to other causes.

The European Union (EU) physical directives on noise and vibration declare three categories of exposure criteria:

	Noise level (dBA)	Acoustic pressure (Pa)
Exposure limit value	87	200
Upper exposure action value	85	200
Lower exposure action value	80	112

Vibration is governed by Statutory Instrument 2005 No. 1093, which has an exemption for air transport which states:

Exemption certificates for air transport 10.–

(1) Subject to paragraph (2), the Executive may, by a certificate in writing, exempt any person or class of persons from regulation 6(4) in respect of whole-body vibration in the case of air transport, where the latest technical advances and the characteristics of the workplace do not

permit compliance with the exposure limit value despite the technical and organisational measures taken, and any such exemption may be granted subject to conditions and to a limit of time and may be revoked by a certificate in writing at any time.

(2) The Executive shall not grant any such exemption unless–

    (a) it consults the employers and the employees or their representatives concerned;

    (b) the resulting risks are reduced to as low a level as is reasonably practicable; and

    (c) the employees concerned are subject to increased health surveil-lance, where such surveillance is appropriate within the meaning of regulation 7(2).

Noise is governed by Statutory Instrument 1989 No. 1790, The Noise at Work Regulations 1989.

5.2 Aircraft Environment

5.2.1 Commercial Aircraft

Noise in modern commercial aircraft has been reduced markedly by leg-islation to reduce noise pollution at airports. The adoption of efficient high-bypass-ratio engines has resulted in most in-service aircraft meeting noise requirements, and this has reduced the noise in the cabin. The cabin occupants therefore are not subject to high levels of audio noise, nor to a great deal of vibration, except for occasional effects of turbulence. Despite this, there is a background persistent sound level and vibration that can contribute to a general annoyance, and may contribute to fatigue on long flights. It is certainly well worth making use of noise-cancelling headphones for those fortunate enough to travel in first or business class, and even the sponge ear plugs make a difference to comfort levels. The House of Lords Select Committee (2000) recommended that airlines should consider offering all passengers the use of simple foam ear plugs.

Noise manifests itself through the following mechanisms as 'background' noise:

- Aerodynamic noise – wind, buffet
- Engine noise and vibration
- Turbulence in poor weather conditions
- Cabin noise, cooling airflow, announcements, passenger noise.

Sound insulation is used in commercial aircraft cabins to reduce noise. The volume of material used is a balance of cost, mass and achieving a target noise level. The material is heavy and takes up space – mass and volume are valuable commodities in any airframe to allow installation of equipment, wiring and ducting. Noise is further reduced by the soft furnishings of the cabin, namely the seats, carpets and curtains.

Noise poses its most severe threat to people involved in aircraft handling when the aircraft is parked on the ramp. Ground crew and those responsible for loading or unloading the cargo bays, servicing or provisioning the aircraft will be issued with ear defenders to protect them. Passengers will be exposed only for short durations on entering or leaving the aircraft, most especially when using steps from the tarmac to the aircraft forward or rear doors. Cabin crew may be exposed for longer periods of time as they wait at the top of the steps to supervise passenger entry or exit. This risk arises from the APU, which is often close to the rear doors, or from ground support vehicle and equipment noise, and from the noise of other aircraft taxiing past.

Internal noise in the cabin can be a source of irritation, although not necessarily a health risk. From the paying passenger's point of view, the most annoying noise can be that made by other people on the flight, and though crews do have the authority to subdue or remove disruptive passengers, there is little they can do to stop babies crying or groups of friends laughing loudly.

That said, in November 2009 US news channels reported that a noisy 2 year old had been removed from a domestic flight with his mother because of safety concerns – other travellers could not hear the pre-flight announcements because of the toddler's excited squeals. The pair later received an apology from the airline.

5.2.2 Military Aircraft

Noise and vibration in the fast-jet military cockpit arise from the combination of a number of sources during normal operation. The noise level in a large reconnaissance type is similar to that encountered on commercial aircraft. Noise levels for personnel carried in a military transport such as a C-130 may be high and accompanied by vibration. These sources are numbered in Figure 5.2 and described below:

1. Aerodynamic noise – wind, buffet.
2. Engine noise and vibration.
3. Turbulence in low-level flight.
4. Cabin noise through the helmet, for example ECS fans, cooling airflow.

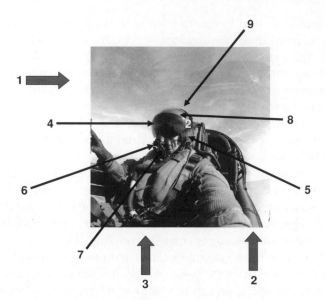

Figure 5.2 Sources of noise and vibration in the military fast-jet cockpit (see text for numbers). Reproduced with permission from BAE Systems.

5. Headset noise – 400 Hz breakthrough and communications static together with incoming speech from other pilots and air traffic control.
6. Microphone and noise transmitted with the comms signal. This includes breathing and transmitted speech, as well as background cockpit noise. This appears in the pilot's own headset as well as being transmitted.
7. Noise transmitted through oxygen mask.
8. Bone-conducted noise through skull to cochlea.
9. Whole-body effect of noise.

In addition there is the possibility of excessive noise in the headset from a failure of the communications system, in which the system may go to full volume. Mission crews in large surveillance aircraft can deal with this by taking off their headsets, but fast-jet pilots cannot remove their helmets. This is an exceptional failure case, but can have a serious short-term impact.

Exposure to excessive noise and vibration leads to a number of issues that affect normal operation, such as difficulty in reading the cockpit displays, tremor in the voice which may distort communications, poor hand–arm co-ordination, and ultimately makes a significant contribution to aircrew fatigue. These issues impose operational limitations but in the long term there is a risk of permanent hearing damage such as tinnitus or loss of hearing at different frequencies resulting from prolonged exposure.

It is likely that many fast-jet aircraft will be unable to meet the current noise limitations without some form of modification. In order to limit exposure without modification to aircraft, a reduction in allowable flying hours for various types will need to be introduced.

5.3 Health Issues

5.3.1 Hearing

Hearing can be affected from frequent and prolonged exposure to loud noise. This often becomes apparent as some form of tinnitus, in which swishing or ringing noises are perceived. This can result in hearing loss in certain frequency ranges or sensitivity to loud noises.

Tinnitus can be apparent at any age, but the condition deteriorates and it is in advanced years that hearing loss is diagnosed. This often occurs when people have retired from the job that caused the condition.

5.3.2 Vibration

Most legislation is aimed at industries in which machine tools are used frequently. Vibration is likely to affect aircraft occupants through whole-body vibration, that is vibration coupled into the airframe from the engines and from turbulence.

The military pilot may experience vibration by keeping hands on the throttles and stick for most of the flight, unlike commercial pilots who mostly fly hands-off. This can lead to a condition known as hand–arm vibration syndrome, which used to be known as vibration white finger. It is a secondary form of Raynaud's disease, in which the affected hands become sensitive to cold.

Patient UK (Patient.co.uk) describes the consequent effects as:

Symptoms may include Raynaud's phenomenon (this is the 'white finger' part), nerve symptoms and muscular aches and pains. Raynaud's phenomenon comes in bouts or 'attacks' that are triggered by cold weather or touching a cold object. A typical bout of Raynaud's phenomenon is as follows:

- At first the fingers go white and cool. This is due to the small blood vessels narrowing (going into spasm).
- They then go a blue-ish colour. This is due to the oxygen being used up from the reduced blood supply of the narrowed blood vessels.
- They then go bright red. This is due to the blood vessels opening up again (dilating) and the return of a good blood flow. This may cause tingling, throbbing and pain.

Some people do not have the full classic colour changes, but still develop bouts of uncomfortable, pale, cold fingers. The duration of each bout of symptoms can last from minutes to hours. The amount of pain or discomfort varies between people. Symptoms usually go after each bout, but one or more blue-ish fingers may persist in the most severe cases.

A Scandinavian study (Burström, Lindberg and Lindgren, 2006) has reported that shock and vibration can cause problems for cabin attendants, particularly during landings. For cabin crew, who generally have seats at the front and rear of the cabin, the dominant direction for vibration load during landing is up and down, although some vibration also occurs in the horizontal. Exposure is worst on the crew seats at the back of the aircraft.

The study also looked at the crew's exposure to shocks and vibration during landing and although it concluded that the risk of damage to health was 'low to moderate', it recommended that efforts be made to prevent injury by developing better seat cushions and back supports and by informing cabin crew about the best posture to adopt for landing.

5.4 System Implications

There are a number of possible alleviations to the issue of noise throughout the system that need to be considered by designers and operators. These include technical, operational and personal solutions.

5.4.1 Limiting Flying Hours

An operational mechanism for limiting exposure can be proposed based on accurate measurement of noise at the ear for each aircraft type. A trial would include installing a test microphone inside the helmet, recording the microphone output during a range of typical mission profiles and analysing the results. This can then be extrapolated to an average exposure over an eight-hour working day to meet the daily dose rate A(8) as defined in the regulations.

This will be difficult for most pilots on operational duties, especially so for qualified flight instructors who need to fly more sorties on training duties than operational pilots.

5.4.2 Active Noise Cancellation

An alternative to limiting flying hours is to reduce the impact of noise at the ear by using noise-cancelling headphones. These reduce unwanted ambient sounds (i.e. acoustic noise) by means of active noise control (ANC). Essentially, this involves using a microphone, placed near the ear, and electronic circuitry which generates a signal in anti-phase with the sound wave arriving at the microphone. This results in destructive interference which cancels out the noise within the enclosed volume of the headphone.

In the domestic world, this technique is used because it is possible to enjoy music without raising the earphone volume excessively. It can also help a passenger sleep in a noisy vehicle such as an airliner, because the technique is especially good at dealing with continuous sounds such as engine noise. The technique is less effective at cancelling high-frequency noise, but good at low-frequency noise, which is what most aircraft cabin noise is.

The use of noise-cancelling headphones in the military aircraft helmet is a feasible proposition, but they must fit into the helmet without compromising its main task of protecting the head, and they must be comfortable.

5.4.3 Microphone Disabling

The mask microphone is known to contribute a lot to noise received in the headset from a combination of breathing, speech and general cockpit noise. Use of the microphone only when essential is a possible solution, as long as it can be designed without degrading the use of the communications system.

5.4.4 Personal Noise Management

Aircrew must be warned of the impact of noise in their daily domestic and office life and asked to restrict their activities to reduce their exposure. Aircrew may wish, for example, to engage in normal domestic activities that contribute to the noise dose. Examples are:

- Driving a noisy car or riding a motorbike.
- Listening to loud music via speakers or iPod.

- Using machine tools for DIY or gardening.
- Engaging in loud motorsport activities.
- Working in a noisy environment – cooling fans, computer fans and so on.

A quiet room may be a suitable solution during working hours.

5.4.5 Risk Assessment

A risk assessment needs to be performed whenever any significant changes to the aircraft configuration of operational procedures are proposed. Examples include the installation of higher rated engines, changes to aerodynamics or changes to operational sortie profiles.

References

Alves-Pereira, M. and Castelo Branco, N. (1999) Vibroacoustic disease: the need for a new attitude towards noise. Centre for Human Performance, Portugal. www.lowertheboom.org (accessed February 2010).

Burström, L., Lindberg, L. and Lindgren, T. (2006) Cabin attendants' exposure to vibration and shocks during landing. *Journal of Sound and Vibration*, **298** (3), 601–685.

House of Lords Select Committee on Science and Technology, 5th Report (15 November 2000) Air Travel and Health, Chapter 1: Summary and Recommendations.

Further Reading

Mansfield, N.J. (2002) Proposed EU Physical Agents Directives on noise and vibration, in *Contemporary Ergonomics* (ed. P.T. McCabe), Taylor & Francis.

Mansfield N.J. (2005). *Human Response to Vibration*. CRC Press, Boca Raton. ISBN 0-415-28239-X.

South, T. (2004) *Managing Noise and Vibration at Work – A Practical Guide to Assessment, Measurement and Control*, Elsevier.

Useful Web Sites

Patient.co.uk

6

Exposure to Radiation

6.1 The Environment

Radiation is all around us, and has been for as long as humans have inhabited the Earth. You do not need to fly to be affected, but can exposure to radiation at altitude be detrimental to health?

Aircraft designers are constrained by practicalities when it comes to minimising the hazards posed by radiation, but good working practises can limit the risk. Even so, over recent years studies have been conducted which suggest that flight crews are more susceptible to certain conditions – including cancer – than those people who spend their entire lives on the ground, and the finger points at radiation as a possible cause.

Radiation exposure can come from the atmosphere and also from the aircraft itself. Take as an example the Shackleton maritime patrol aircraft, where the radar operator sat at a workstation in close proximity to the klystron oscillator, which provided the radar energy. A number of operators are known to have suffered from testicular cancer, which may be attributable to this 'close contact' radiation. One radar operator has revealed (author interview) that he had developed testicular cancer and needed surgery and said that he knew of a number of his colleagues who were similarly afflicted. The location of the radar components is shown in Figure 6.1 for the AEW and MR versions of the Shackleton.

But the Shackleton is not the only aircraft to come under the spotlight because of possible cancer links. There is also concern that female aircrew

Air Travel and Health: A Systems Perspective   Allan Seabridge and Shirley Morgan
© 2010 John Wiley & Sons, Ltd

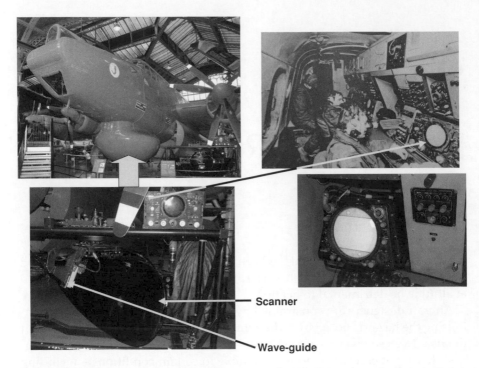

Scanner

Wave-guide

Figure 6.1 Radar components in Shackleton aircraft. Reproduced with permission from Allan Seabridge.

flying long hours on commercial aircraft may be more likely to suffer from breast cancer than women in the general non-flying population.

Furthermore, an increase in testicular cancer has been noted in police officers using hand-held traffic radar guns (Davis and Mostofi, 1993).

Aircraft and aircrew can be subject to ionising and non-ionising radiation throughout their flight envelope. The sources of radiation can be natural, as in cosmic or UV radiation, or human-made, as in radio frequency (RF) radiation from radar and microwave devices.

6.1.1 Cosmic Radiation

Cosmic radiation is a natural phenomenon for which there are no practical engineering solutions to prevent its impact on aircrew. The particles involved contain too much energy to be absorbed by materials that are suitable for use in aircraft construction – lead or concrete would provide some protection, but are clearly not suitable materials for aircraft manufacture. This naturally occurring 'background' ionising radiation affects everyone on Earth, but people have been deemed to be especially susceptible if they spend significant periods of time above 26 000 ft – or FL260 in aviation terminology.

The WHO describes cosmic radiation as 'part of our natural environment and we are constantly exposed to a certain amount of ionizing radiation. Radiation originating from outer space and the sun is called cosmic radiation and contributes about 13% of the background radiation level on Earth (a greater part is due to radon)' (who.int).

Robert Barish, physicist and author of *The Invisible Passenger: Radiation Risks for People Who Fly* (1996), reminds us that the Sun is simply a big thermonuclear device and believes airline crew members are exposing themselves to more radiation than almost any other occupation.

Writer Bill Bryson puts it even more colourfully:

Cosmic background radiation is something we have all experienced. Tune your television to any channel it doesn't receive and about 1 per cent of the dancing static you see is accounted for by this ancient remnant of the Big Bang. The next time you complain that there is nothing on, remember that you can always watch the birth of the universe.

(Bryson, 2003)

Cosmic radiation is a complex mixture of both charged and neutral particles, some of them generated when primary particles from space interact with the Earth's atmosphere, which acts as a radiation shield.

High-energy cosmic rays bombard us all the time, but because they interact quickly with our atmosphere they produce particles of much lower energy, which hit the Earth harmlessly. In terms of human exposure, one feature of cosmic radiation is of particular importance: most of the effective radiation dose from cosmic radiation is due to neutrons of different energy levels. Neutrons are subatomic particles which – when compared with X-rays or gamma rays – potentially cause more biological damage per dose unit.

Because cosmic rays are made up of many different types of radiation from a wide range of energies it can be hard to measure radiation doses, and sophisticated approaches are required to deal with this.

The WHO summarises the current knowledge and ongoing research activities, and gives target-group specific recommendations on actions concerning cosmic radiation.

The International Agency for Research on Cancer (IARC) has concluded that there is sufficient evidence that neutrons – which constitute 30 to 60% of cosmic radiation – are carcinogenic to humans (Rafnsson *et al.*, 2005).

6.1.2 Radiation in Space

The effect of radiation in space was certainly of concern to the US space mission medical teams. Because astronauts were operating for prolonged periods beyond the Earth's atmosphere, they were instructed to wear radiation monitoring badges, and missions were timed to avoid dangerous periods of solar activity. On Apollo 11 the astronauts noticed flashes in the cabin. According to Buzz Aldrin there were 'little flashes inside the darkened cabin, spaced a couple of minutes apart'. He also noticed double flashes. The other astronauts also noticed the effect, on one occasion counting over 50 flashes in an hour (Parry, 2009).

It was at first believed that the flashes were taking place in the cabin, but the theory emerged that they were occurring in the eyeball. These were cosmic rays passing through the eye and appearing as a flash of energy. This phenomenon has also been reported by astronauts aboard Skylab, the Space Shuttle, the International Space Station and Mir. The phenomenon will become of more interest as the possibility of space tourism becomes realistic.

6.1.3 Non-ionising Radiation

Radiation from RF sources can also be a risk to humans. This radiation is non-ionising, but nevertheless can damage human tissue, and legislation exists to limit exposure. Radiation from radars, whether on the aircraft in question, from close-flying aircraft or from high-energy ground-based radars, are all potential threats. There are groups of people who claim to suffer from WiFi installations or from living in close proximity to overhead power lines and there is a growing body of work that suggests that excessive use of mobile phones may also be hazardous.

ICNIRP is the International Commission on Non-Ionizing Radiation Protection. It is a body of independent scientific experts consisting of a main commission of 14 members, four scientific standing committees covering epidemiology, biology, dosimetry and optical radiation, and a number of consulting experts. This expertise is brought to bear on addressing the important issues of possible adverse effects on human health of exposure to non-ionising radiation.

The guidelines produced by ICNIRP are very likely to become mandated under law across the EU by 2012. These will provide guidelines on electric and magnetic field strength exposures for the general public, both recommended and peak limits, and also occupational limits. The occupational limits differ for the general public because workers exposed to known hazards will have their exposure monitored and future exposure can be limited by working practices.

6.2 The Aircraft Environment

6.2.1 Ionising Radiation from Space

Cosmic radiation is the collective term for the radiation which comes from the Sun (the solar component) and the galaxies in the Universe (the galactic component). Cosmic radiation is ionising: that is, it can displace particles from atoms. This can lead to the disruption of molecules in living cells (Hunter, 2003) and this type of cell damage has been identified as a contributory cause of cancer. Certain cells are considered to be more susceptible to radiation damage than others – bone marrow, the lining of the intestine and hair follicles are particularly vulnerable (this is the reason that radiation therapy as a treatment for cancer can produce nausea, anaemia and hair loss). It has been estimated that the chance of a fatal cancer occurring would be approximately 1% following 30 years of commercial flying at 1000 hours per year, hence the risk is small. However, small doses of radiation are known to start off chains of events that may lead to the occurrence of cancer after an interval of many years.

Some studies have been carried out on commercial aircrew (Blettner *et al.*, 2003a; Whelan, 2003; Blettner *et al.*, 2003b). Epidemiological studies of pilots and cabin crew suggest that pilots are at an increased risk of malignant melanoma skin cancers. There are also indications of raised risk of breast cancer amongst female cabin crew, but it is not clear whether these observed excesses are due to occupational cosmic radiation or to lifestyle, reproductive or other non-occupational factors. At present, there are some indications of a possible increased cancer risk in commercial aircrew and current US and European studies would provide a better understanding (Clarke, 2003).

Flying at high altitude exposes aircrew to ionising radiation arising from solar events. This is a well-understood phenomenon which is regulated in the nuclear industry. During the major solar radiation storms of 2003, some transatlantic flights were held on the ground for several hours until the storms subsided. This disrupted some airline and space satellite communications and led some electrical grids to curb their power transmissions as a precaution. The storms were triggered by giant eruptions on the Sun's surface, known as solar flares, and were powerful enough to disable temporarily a Japanese satellite. Such activity can be harmful to astronauts, hence can affect prospective space tourists, and contributes to the intensity and variability of cosmic radiation in the Earth's atmosphere.

Not surprisingly, the behaviour of solar storms is of great scientific interest. A recent research activity has been launched to engage widespread public support to identify and measure solar storms as part of the Zooniverse internet activity. Information can be found on solarstormwatch.com.

Legislation is in place to measure the impact on regular, high-altitude-flying aircrew in the commercial world. The situation is less well regulated in the military flying community. Individual monitoring is to be regarded as best practice when it comes to looking at radiation and its effects on frequent flyers, but it should be recognised that this can impose unjustifiable cost for some operators.

Cosmic radiation exposure occurs as a result of long-duration, high-altitude flight (CAP or long transit/ferry) with full transparency canopy, or frequent transitions to very high altitude.

When cosmic rays consisting of energetic particles originating from outer space, or the flow of fast-moving charged particles in the solar wind from the Sun, collide with the Earth's atmosphere, they produce a cascade of lighter particles, a so-called 'air shower'. High-energy neutrons from air showers collide with microchips and upset or damage microelectronic devices. These occurrences, known as 'single-event effects' (SEEs), can affect circuitry on the ground, but the problem is 300 times greater at high altitude. This makes it of particular concern to both the civil and military aerospace industries. Although this effect has been recognised for some time, it is particularly relevant now because of the demand for greater memory (RAM) density in computers. Smaller electronic circuitry is more vulnerable to this buffeting from neutrons.

A microchip in an aircraft can be struck by a cosmic neutron every few seconds. When a neutron hits silicon, a nuclear reaction occurs, causing an electrical charge shower that can interfere with the normal operation of electronic equipment. This can lead to temporary loss of RAM or even permanent burnout of equipment, but most equipment or system design strategies will enable such transient faults to be absorbed without significant performance degradation.

6.2.2 Non-ionising RF Radiation

RF exposure in flight can occur while flying in formation with other aircraft using radar. Tanker aircraft crew may also be illuminated by radar if the recipient aircraft continues to transmit on approach to the drogue as

Figure 6.2 Tanker and receiving aircraft. Reproduced with permission of U.S. Air Force photo/Staff Sgt. James L. Harper Jr, af.mil. See Plate 12 for the colour figure.

shown in Figure 6.2. This exposure can be reduced or eliminated by adherence to procedures calling for the inhibition of radar transmission during such operations.

There is a risk of radiation to aircrew from the radar sidelobes or back lobes, especially in a fighter aircraft where the cockpit is located immediately aft of the radar scanner. The main beam of a radar is central to the performance of the system and is used to direct energy at a target, from which energy is returned to the scanner. However, the radar is prone to produce sidelobes of energy as illustrated in Figure 6.3, and these detract from the performance of the radar. In the worst case these lobes may actually be directed backwards as back lobes and at extreme angles of the scanner may illuminate the pilot of a fighter jet, thereby causing the pilot to be radiated with non-ionising RF energy.

From March 2004 the National Radiological Protection Board (NRPB) recommended that the UK adopt the guidelines of the ICNIRP for limiting exposures to electromagnetic fields between 0 and 300 GHz. The guidance is more restrictive than previous recommendations and work will have

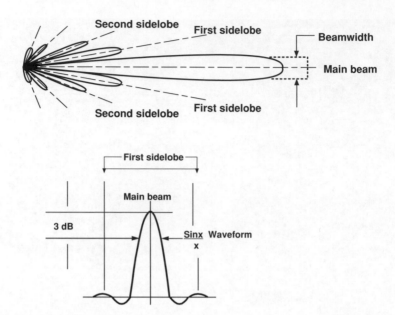

Figure 6.3 Illustration of radar main lobe and sidelobes. Reproduced with permission from Moir & Seabridge (2006).

to be done to understand the impact on aircrew and ground crews. Note that the NRPB has now been absorbed by the Health Protection Agency (www.hpa.org.uk). The issue of the EU Physical Agents Directive on exposure of workers to electro-magnetic radiation has been postponed until 2012.

6.2.2.1 Radiation Beyond the Control of Designers

There are, however, sources of exposure to non-ionising radiation that are beyond the control of aircraft designers. There have been reports for some time from people who claim to suffer from RF radiation from sources such as mobile phones, overhead power lines and WiFi installations, and since the majority of airline passengers and crew are exposed to this potential threat every day of their lives, then it is difficult to separate this risk from the risk of on-board sources of radiation.

On the subject of mobile phones the *Daily Telegraph* (Lean, 2009) reported on the subject of mobiles and cancer:

Does use of mobile telephones lead to cancer? There seems to be no resolution to this question. Partly this is because cancer and associated tumours take some time to develop, and the answer may lie sometime in the future. However, evidence seems to be increasing, especially for children and people who have used handsets for more than 10 years. Since the mobile handset is now ubiquitous, and is becoming the main means of telephonic communication, the sample of users has increased since the early suspicions were voiced, with more than 2 billion users worldwide.

The report goes on to declare that a Swedish study of long-term users indicates on average that they are about twice as likely to get malignant gliomas (an incurable brain cancer) on the side of the head where they hold their handset. More Swedish research found that people who started to use mobiles before the age of 20 were five times more likely to contract cancers, and eight times more prone to get them on the relevant side of the head.

Sufferers from WiFi sensitivity claim symptoms such as headaches, nausea, stomach upsets, tinnitus, brain fog, skin sensitivity and short-term memory loss. Not surprisingly these claims have been contested on the basis that the electro-magnetic field induced by WiFi transmissions is not sufficient to cause problems to humans.

Since the modern aircraft contains many transmitters and sources of relatively high-frequency energy, there may be some combinatory risk. Systems designers take care to limit the exposure of crew and passengers, abiding by statutory regulations limiting the energy of radiated field strength.

Further protection is afforded by the fact that the sources of RF power are located remotely from passengers. All relevant cables are screened and the transmitting and receiving antennas are installed on the exterior of the aircraft. All appropriate electro-magnetic health (EMH) measures will have been applied during design to avoid mutual systems interference and RF losses.

This is a quite different situation, much more controlled, compared with that which the general public are exposed to on the ground. In the ground case hand-held radios and mobile phones are used frequently and held close to the head. There is more discussion on this subject in Chapter 12.

6.2.3 Understanding the Risk in Commercial Aircraft

A joint study group has conducted measurements on a number of long-haul commercial aircraft operations. The study was sponsored by the Particle Physics and Astronomy Research Council (PPARC), now known as the

Science Technologies and Facilities Council. The study team included members from the following organisations:

- Mullard Space Science Laboratories (part of University College London)
- Virgin Atlantic Airways
- The CAA
- The National Physical Laboratory.

This group published work that showed mean dose measurements from long-haul flights on routes operated by Virgin Atlantic Airways. The measurements were made by a tissue equivalent proportional counter (TEPC) designed and made by SolarMetrics. The TEPC was designed to mimic human tissue and to provide a measure of the dose equivalent to a few microns of tissue. Measurements were recorded every minute during each flight and later integrated with the aircraft's flight data recorder to include time, latitude, longitude and flight level (Iles *et al.* 2003; Taylor *et al.*, 2002). The measurements were able to show variations in dose rate with latitude and solar activity.

Results of measurements from 13 different routes were produced (Taylor *et al.*, 2002) which include routes between London and Athens, Boston, Chicago, Glasgow, Hong Kong, Johannesburg, Los Angeles, Miami, New York, Orlando, Shanghai, San Francisco and Tokyo. The results provided a measured dose for each leg of each flight. Factoring this dose for flight duration indicates that a dose of 4 µSv per hour can be used as an approximate measure of expected dose for each hour spent above 26 000 ft.

British Airways has estimated the flight exposure in order to monitor its crews and produced the following indicators (britishairways.com):

Concorde	12 to 15 µSv per hour
Long-haul aircraft	5 µSv per hour
Short-haul aircraft	1 to 3 µSv per hour

The Concorde average dose rate was high because of the altitude at which it flew when in service; however, the speed of its flight meant that exposure was of significantly shorter duration.

Table 6.1 shows estimated radiation doses for selected flights issued by the WHO, which are indicative values only, based on a cruise altitude of 10 000 m (33 000 ft) (calculations from http://www0.gsf.de/epcard2/index_en.phtml, 30 November 2005). The natural background radiation amounts to 2 to 3 mSv per year at most geographical locations worldwide (who.int).

Table 6.1 Cosmic radiation dose on selected flights (www.who.org).

From	To	Duration (hours)	Estimated radiation dose (µSv per hour)
Sydney	Singapore	7.5	17
Bangkok	Washington via Sydney, Los Angeles	28	70
London	Tokyo	12	58
Buenos Aires	Athens	18.35	41
New York	Paris	7	35
Frankfurt	Los Angeles	9.5	51
Johannesburg	Mumbai	9.1	16

Assessments of individuals' exposures may be made using dose estimates for routes calculated using a computer program, combined with staff roster information. The data to be input are: the date of departure, the location of departure, the flight profile detailing the time in ascent, cruise and descent and the arrival location.

There are several simple-to-use programs which have been validated. One is produced by the Civil Aerospace Medical Institute in the US (formerly known as the Civil Aeromedical Institute), the latest version being CARI-6M. This is freely available from the US Federal Aviation Administration and can be downloaded at www.faa.gov/data research/research/med human facs/aeromedical/radiobiology/cari6.

Table 6.1 shows some variability between flights and does not always agree with the figures above, that is 5 µSv per hour for long-haul travel. This may be explained by the variability of cosmic ray intensity with latitude, and also solar activity. This means that the exposure of any one aircraft will vary throughout its route, as well as from route to route, whether great circle or transpolar.

6.2.4 Understanding the Risk in Military Aircraft

The research above seems to indicate that a suitable dose rate for comparison of exposure in different aircraft types is 5 µSv per hour for each hour spent at or above 26 000 ft. The trigger point at which monitoring of aircrew dose rate is required is 1 Sv per year in accordance with the ANO. This equates to 200 hours or more per year above 26 000 ft.

For the purpose of calculating the risk to aircrew a number of aircraft types were examined to determine how much time each type was likely to spend above 26 000 ft each year in normal operations. The following types were selected:

Fast jet – the sortie profiles do not include significant durations at 26 000 ft, other than CAP, and the sortie duration is short. It is unlikely that fast-jet pilots will exceed 200 hours above 26 000 ft.

Trainers – primary jet trainers such as the BAE Systems Hawk spend most of their time at low altitude and it is unlikely that pilots or students will exceed 200 hours above 26 000 ft.

Large reconnaissance – aircraft such as maritime reconnaissance, battlefield surveillance and Airborne Early Warning (AEW) aircraft have long-duration missions. Maritime patrol includes significant time at low level looking for surface or subsurface assets. However, the other two types may spend significant time at high altitudes in order to obtain maximum ground coverage or early air target identification.

Transport – Materiel and troop transport aircraft may spend long durations at altitude to and from war zones. The crews are likely to be at risk.

Tankers – in-flight refuelling aircraft may spend long durations at altitude during tanking operations, as well transiting to and from the in-flight refuelling rendezvous points. The crews are likely to be at risk, although their deployments may be irregular.

High-altitude photo-reconnaissance – may fly at significantly higher altitudes than most types and may be at risk for much of their missions. However, the missions may be intermittent and of relatively short duration.

Rotary wing – spend most of their time at low altitude and it is unlikely that pilots or aircrew will exceed 200 hours above 26 000 ft.

Table 6.2 summarises the level of risk for each type.

6.3 Aircraft Systems

There is very little impact on aircraft systems from radiation, other than semiconductor transient damage noted above. The random nature of the effect means that practical monitoring cannot be done effectively, and the high energy of the particles means that shielding of passengers or crew is impossible. Careful monitoring of crew rosters is required to ensure that the permitted dose level is not exceeded. The use of models to do this is highly recommended.

Plate 1 Travel sweets were an early attempt to overcome the effects of cabin pressure altitude on the ear canal (A.L. Simpkin).

Plate 2 The airport concourse, where many travellers first experience anxiety from security checks, baggage check-in, potential delays and long waits.

Plate 3 A study in contrasts: the Airbus A340 (top) as an example of a twin-aisle aircraft and the Boeing 737 (bottom) as an example of a single-aisle aircraft.

Plate 4 Figure 2.1 Young children and babies are prone to changes in pressure and are too young to understand how to react to the situation. Their response may disturb other passengers and cause the parent more stress.

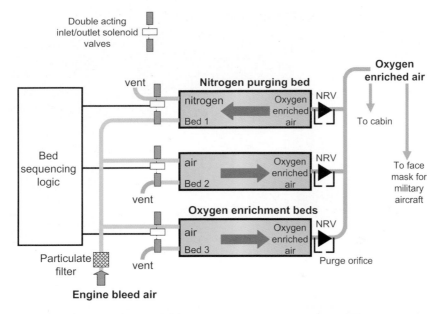

Plate 5 Figure 3.5 OBOGS: on-board oxygen generation.

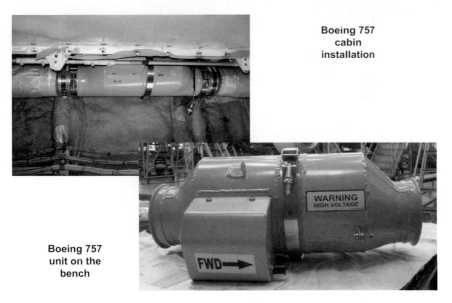

Plate 6 Figure 3.11 The AirManager filtration system developed by Quest International and BAE Systems to improve the quality of cabin air.

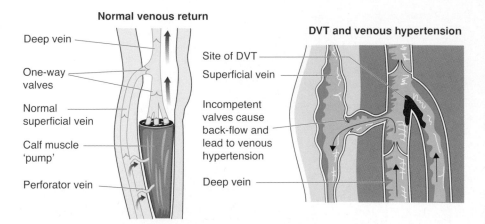

Plate 7 Figure 4.1 Venous return showing the potential sites for deep vein thrombosis.

Plate 8 Figure 4.3 Examples of passenger cabins for single- and twin-aisle types.

Plate 9   Figures 4.4 and 7.4 Aircraft seats showing the range of seating styles in different cabins.

Plate 10   The Virgin Atlantic upper-class seat which turns into a bed at the press of a button and gives the traveller a private pod.

Plate 11 Figure 5.1 The Avro Shackleton, a source of hearing-related issues for many aircrew.

Plate 12 Figure 6.2 Aircraft receiving fuel from an airborne tanker may inadvertently expose the tanker crew to non-ionising radiation.

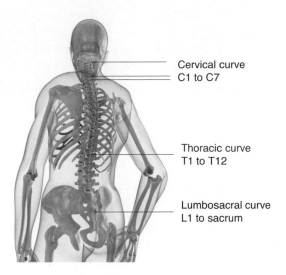

Cervical curve
C1 to C7

Thoracic curve
T1 to T12

Lumbosacral curve
L1 to sacrum

Plate 13   Figure 7.1 The human spine – the source of lower back pain.

Plate 14   Figure 8.2 Typhoon aircrew in full flying gear.

Plate 15 Figure 9.2 Some examples of workstations where mission crew members may spend many hours doing a high vigilance task in arduous conditions.

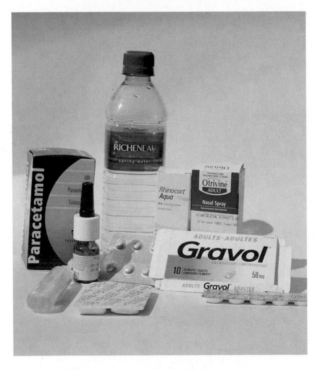

Plate 16 A range of prescription and proprietary medicines is available to treat symptoms – but medical advice should always be sought.

Table 6.2 Summary of risk for military types.

Type	High risk	Medium risk	Low risk
Fast jet			Y
Trainer			Y
Large reconnaissance	Y		
Transport	Y		
Tanker		Y	
Photo-reconnaissance		Y	
Rotary			Y

6.4 Health Issues

6.4.1 Risk of Cancer

Airline pilots and astronauts have been found to have an increased risk of mutations to genes in their blood cells and certain cancers linked to cosmic radiation (BBC News, 'Flying boosts radiation dose', 10 December 1999). A recent study found an increased rate of chromosomal translocations in airline pilots with long-term flying experience. For each one-year increase in flight years, the likelihood of such a translocation rose by 6% (Yong *et al.*, 2009). Translocation is commonly seen in cancer and is indicative of cumulative ionizing radiation exposure.

Several studies have found a link between air travel and cancer. These studies, however, involved people who fly for a living, that is professional pilots, cabin crew and military mission crews. The findings consistently show increased risks of breast cancer amongst female flight attendants and skin cancer amongst both pilots and flight staff of commercial aircraft (Sigurdson and Ron, 2004). Exposure rates for professional pilots and aircrew may be many times greater than for people at ground level (Barrish, 1999).

In a British Airways survey of 411 airline pilot deaths, incidences of malignant melanoma, colon and brain cancers were observed to be slightly higher than normal (Blettner *et al.*, 2003a,b). A large study of Nordic airline pilots concluded that the pilots assessed did have an increased risk of skin cancer (Pukkala *et al.*, 2003). The study, however, could not attribute the rate to cosmic radiation alone, but equally could not rule it out as a contributing factor. A Norwegian study found a statistically significant exposure–response relationship between the cumulative radiation dose and malignant melanoma (Haldorsen, Reitan and Tuelen, 2000).

But the overall picture remains unclear. A study of Swedish pilots showed overall cancer incidence to be similar to the Swedish male general population (Hammar *et al.*, 2002). An increased incidence of malignant melanoma in airline pilots was reported and other skin cancers in military pilots, but this could be associated with exposure to UV radiation either at work or outside of work. A study of mortality and life expectancy of male British Airways flight-deck crews and the occurrence of cancers of the brain, central nervous system, colon and melanoma concluded that flight-deck crews actually *live longer* than the general England and Wales population and do not exhibit patterns of death that could be directly attributable to their occupation (Irvine and Davies, 1999).

Researchers have uncovered an excess of other cancers amongst pilots, although the studies are inconsistent. One found a significantly higher risk of acute myeloid leukaemia (AML), whilst others showed higher rates of cancers of the colon, brain, prostate and rectum and Hodgkin's disease (Gundestrup and Storm, 1999; Rafnsson, Hrafnkelsson and Tulinius, 2000). The same researchers who found the link between pilots and AML (see above) also discovered evidence that cosmic radiation may well be the culprit. They found chromosomal damage in 57% of flight staff with AML or myelodysplasia (a 'pre-leukaemia' bone-marrow disorder) compared with only 11 per cent in non-flight staff. The researchers concluded that the chromosomal abnormalities seen in myelodysplasia and AML could indicate 'previous exposure to ionizing radiation' (Gundestrup *et al.*, 2000).

As for skin cancer (malignant melanoma), incidence rates for cabin crew are two to three times the expected rate (Linnersjö *et al.*, 2003; Osamu Tokumaru *et al.*, 2006), while commercial airline pilots may be 10 times more likely to develop melanomas compared with the general population. Some pilots, such as those on international routes, appear to have an even greater risk (Rafnsson, Hrafnkelsson and Tulinius, 2000).

But is it the job and the inherent radiation risk that threatens pilots and cabin staff or associated lifestyle factors? Increased rates of breast cancer might be attributed to reproductive factors, since female cabin crew often have no children, or have them at a later age than average, both of which are risk factors for breast cancer. Higher skin cancer rates could be due to more exposure to the Sun, since aircrew may be spending more time in sunnier destinations than the general population. (Lim and Bagshaw, 2002).

However, several studies have taken these factors into account and concluded that they are unlikely to explain all of the increased risk of cancer amongst flight crew (Linnersjö *et al.*, 2003). As some scientists point out, the irregular working hours of flight crew and frequent disturbance of the circadian rhythm (jet lag) may also play a role in increasing cancer risk (Whelan, 2003). Another factor is that the inhabitants of aircraft are subject to a number of contributory factors simultaneously, not merely singly. This combined effect, where the aircraft acts as an integrating mechanism, may increase susceptibility to environmental effects.

6.4.2 Risks to Female Crew Members

There may be a higher incidence of bone and breast cancer amongst female cabin crew (Ballard *et al.*, 2000). In one study from Japan, female flight attendants had a 40% increase in breast cancer compared with the general population (Osamu Tokumaru *et al.*, 2006). Other studies from Finland, Iceland, Norway, Denmark and the US also found higher breast cancer rates in female cabin crew, up to five times higher in those with five years or more on the job (Rafnsson *et al.*, 2003a).

There is also evidence that cosmic radiation is dangerous to the foetuses of pregnant flight staff, particularly between 8 and 15 weeks of pregnancy. According to the US Air Force Inspection Agency, both animal and human studies of ionising radiation exposures suggest that pregnant flyers may subject their foetuses to a risk of decreased cognitive capacity as well as childhood leukaemia (Geeze, 1998). For this reason, flight staff are advised to limit air travel during pregnancy, but may still unknowingly exceed the recommended foetal dose limit of 1 mSv over the entire pregnancy (www.who.int/ionizing_radiation/env/cosmic/WHO_Info_Sheet_Cosmic_Radiation.pdf). A recent study found that the dose to the foetus can exceed 1 mSv after 10 round trips on commercial flights between Toronto, Canada, and Frankfurt, Germany (Chen and Mares, 2008).

6.4.3 Risks to Passengers

Are there radiation risks to passengers, as well as crew members? Some studies consider that frequent flyers are at risk from increased doses of radiation, and the WHO advises those concerned about cosmic radiation to:

- keep informed about the health effects of cosmic radiation;
- limit air travel during pregnancy.

Frequent flyers whose air time is similar to that of aircrew are advised to record their personal cumulative radiation doses on a regular and permanent basis and consider radiation exposures when selecting flight schedules.

6.4.4 Eye Damage

There is evidence to suggest that cosmic radiation may cause damage to proteins in the lens of the eye. Commercial airline pilots are reported to be three times more likely than normal to develop cataracts (BBC News, 'Cosmic rays harm pilot's sight', 8 August 2005). Furthermore, the increased exposure is reported to be related to an increased probability of developing nuclear cataracts (BBC News, 'Cosmic rays harm pilot's sight', 8 August 2005; BBC News, 'Airline threat to be assessed', 9 August 2000).

A University of Iceland study found that commercial pilots were three times more likely than usual to develop cataracts and noted that the increased risk could not be explained by other factors such as UV exposure and smoking, known to be risk factors. It concluded that 'cosmic radiation may be a causative factor in nuclear cataracts amongst commercial airline pilots' (Rafnsson et al., 2003b). Such an increased risk of cataracts has also been reported by the US National Cancer Institute in radiological staff exposed to ionising radiation (Chodick et al., 2008).

A high proportion of astronauts developed cataracts in later life, which could have been caused by exposure to cosmic rays from traversing the Van Allen Belt (about 30 minutes) and spending time in deep space (Stroud, 2009). It has been reported (Cucinotta, 2001) that 39 former astronauts have been affected and 36 of these had flown high-radiation, deep-space missions such as the Apollo missions. Some cataracts developed 4 to 5 years after the mission, whereas others took 10 years or more to develop.

6.4.5 Damage to Life Support Equipment

There is a potential threat, albeit indirect, to people who rely on electronic equipment to protect them from medical conditions if an SEE disrupts the equipment. Devices such as heart rate monitors or pacemakers may cause immediate discomfort, whereas devices such as blood-sugar-level monitors may give an imprecise or erroneous reading when used, though, for most people, the airport security systems are likely to cause a problem, not the flight itself.

6.5 System Implications

6.5.1 Legislation

6.5.1.1 EU Council Directive 96/29

EU Council Directive 96/29 Euratom provides advice on basic safety standards for the protection of the health of workers and the general public against the dangers arising from ionising radiation. Article 42 of the Directive is titled 'Protection of Air Crew' and states:

> Each member state shall make arrangements for undertakings operating aircraft to take account of exposure to cosmic radiation of aircrew who are liable to be subject to exposure to more than1 milliSievert (mSv) per year. The undertakings shall take appropriate measures, in particular:

> - To assess the exposure of the crew concerned.
> - To take into account the assessed exposure when organising working schedules with a view to reducing doses of highly exposed aircrew.
> - To inform the workers concerned of the health risks their work involves.
> - To apply Article 10 for female aircrew.

Article 10 refers to pregnant women and delegation to the unborn child of rights of protection equivalent to members of the public. This means that, once the pregnancy is declared, the female crew member must plan for future occupational exposures to the foetus to be not greater than 1 mSv during the remainder of the pregnancy. This may be applicable to female aircrew and mission crew of long-range surveillance aircraft.

The 1 mSv trigger level is considered likely to affect all aircrew of aircraft which regularly operate above 26 000 ft. Aircraft operators are required to assess the exposure to cosmic radiation of aircrew who are liable to be subject to radiation in excess of this trigger level.

6.5.1.2 Air Navigation Order 2000

The relevant content of directive 96/29 has been incorporated into Air Navigation Order 2000. The following text has been extracted from ANO (2000) Part VI – Fatigue of Crew and Protection of Crew from Cosmic Radiation. Especially relevant items have been reproduced in bold text:

Application and interpretation of Part VI
71-(1)

(a) Subject to sub-paragraph (b) **articles 72 and 73** of this Order apply in relation to any aircraft registered in the United Kingdom which is either:
  (i) engaged on a flight for the purpose of public transport; or
  (ii) operated by an air transport undertaking.

(b) Articles 72 and 73 of this order shall not apply in relation to a flight made only for the purpose of instruction in flying given by or on behalf of a flying club or flying school, or a person who is not an air transport undertaking.

71-(2) For the purposes of this Order:

(a) 'flight time', in relation to any purpose, means all time spent by that person in:

  (i) a civil aircraft whether or not registered in the United Kingdom (other than such an aircraft of which the maximum total weight authorised does not exceed 1600 Kg and which is not flying for the purpose of public transport or aerial work), or:

  (ii) a military aircraft (other than such an aircraft of which the maximum total weight authorised does not exceed 1600 Kg and which is flying on a military experience flight),

while it is in flight and he is carried therein as a member of the crew thereof.
Protection of aircrew from cosmic radiation.
    75-(1). A relevant undertaking shall take appropriate measures to:

(a) **assess the exposure to cosmic radiation when in flight of those aircrew who are liable to be subject to cosmic radiation in excess of 1 mSv per year;**
(b) take into account the assessed exposure when organising work schedules with a view to reducing the doses of highly exposed aircrew; and
(c) inform the workers concerned of the health risks their work involves.

75-(2). A relevant undertaking shall ensure that in relation to a pregnant aircrew member, the conditions of exposure to cosmic radiation when she is in flight are such that the equivalent dose will be as low as reasonably achievable and is unlikely to exceed 1 mSv during the remainder of the pregnancy.
    75-(3). Nothing in paragraph 75-(2) shall require the undertaking concerned to take any action in relation to an aircrew member until she has notified the undertaking in writing that she is pregnant.
    75-(5) In this article and in article 77:

(a) 'aircrew' has the same meaning as in article 42 of Council Directive 96/29 Euratom of 13th May 1996, and

(b) 'undertaking' includes a natural or legal person and 'relevant undertaking' means an undertaking established in the United Kingdom which operates aircraft.

75-(6). In this article:

(a) 'highly exposed aircrew' and 'milliSievert' have the same respective meanings as in article 42 of Council Directive 96/29 Euratom of 13th May 1996, and
(b) 'year' means any period of twelve months.

6.5.1.3 Interpretation of Legislation

Although it is tempting to regard Directive 96/29 as being applicable only to commercial aircraft, the implementation of Article 42 makes it clear in paragraph 8 that 'the requirement will affect any undertaking operating aircraft and not just public transport operators'. Similarly, it should be noted that Article 71 of the ANO applies the terms of 'public transport' or an 'air transport undertaking' to Articles 72 and 73 only. In Article 75, specifically addressing radiation, the ANO refers only to 'undertakings'.

It can be inferred from this that operation of military aircraft is not exempt from legislation. Operators and manufacturers should apply the legislation to their operations as flight test operations where relevant in the employment and tasking of aircrew. It can also be inferred that manufacturers have an obligation to inform users of their products that they should be aware of legislation and how it applies to their aircrew.

JSP 553 is a major regulatory instrument that governs the operations of UK aircraft manufacturers. The JSP makes it clear that the majority of the ANO is not relevant to manufacturers' operations, except that Paragraph 71(2)(a) is applicable. The JSP, however, does state:

1.5. Notwithstanding the fact that the majority of the provisions of the ANO do not apply to military aircraft, the Crown could be liable in common law if it were to operate its aircraft negligently and cause injury or damage to property. It is therefore the view of the MoD legal advisors that internal MoD regulatory arrangements should be at least as effective as those in respect of civil aircraft contained in the ANO.

6.5.2 Mitigating Action

For those who fly as a career, it is clear that mitigating action should be taken to reduce the possible risks caused by radiation.

The International Federation of Airline Pilots Associations (IFALPA) recognises 20 mSv per year as the cosmic radiation limit for airline flight crews as established by the National Council on Radiation Protection and Euratom.

Adapting aircrew rosters to ensure that exposure is maintained within tight and known limits is the ideal approach. This entails a detailed monitoring regime for personal flying hours and routes, and a process should be in place to demonstrate that this is being done.

Aircrew fitness and long-term good health are paramount in minimising the effects of flight-induced health risks. There are suggestions that the effect of regular medicals and better levels of general fitness may actually be disguising or outweighing the increased risk of cancer in pilots (BBC News, 'Flying boosts radiation dose', 10 December 1999).

Protection from non-ionising radiation is best effected by good practice in the use of radar. This includes the posting of notices warning of the source of such radiation, notes in the pilot's manual prohibiting the use of radar in prescribed circumstances, and the provision of radar absorbent material (RAM) on bulkheads between the radar scanner bay and the cockpit, as illustrated in Figure 6.4.

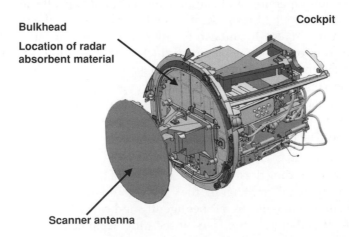

Figure 6.4 Radar absorbent material on bulkhead. Reproduced with permission from BAE Systems.

6.6 Future Developments

Virgin has recently unveiled its plans for space tourism using a two-stage vehicle designed by Bert Rutan (virgingalactic.com). The vehicle consists of a twin-fuselage mother-ship carrying a small spacecraft capable of flying into suborbital space.

The plan is for the mother-ship to climb to 50 000 ft and release the tourist-carrying spacecraft. A hybrid rocket engine will lift this craft to about 360 000 ft. After a short period of suborbital flight, including a weightless experience, the craft will return to its landing ground. An indication of the space profile is shown in Figure 6.5.

This flight profile poses two specific risks to passengers and crew:

1. The crew and passengers of the mother-ship will be subject to radiation between 26 000 and 50 000 feet. The crew can be expected to experience this on a regular basis and should have their exposure logged.

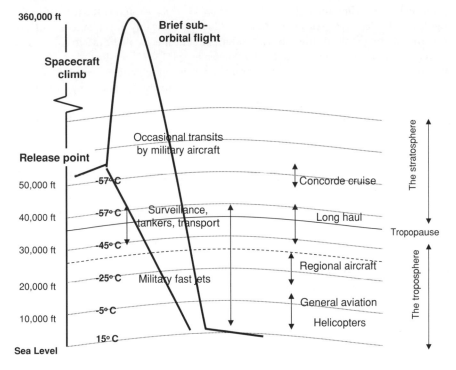

Figure 6.5   An indication of the Virgin Galactic flight profile.

2. The crew and passengers of the spacecraft will be subject to enhanced levels of radiation during their trip to space and back. The exposure will be short for passengers, although in excess of what can be expected from a high-altitude commercial flight. For the crew this exposure will be repeated, and their exposure will need to be logged. The levels of radiation and the altitude may well be beyond the capability of the current CARI computer model.

References

Ballard T. *et al.* (2000) Cancer incidence and mortality among flight personnel: a meta analysis. *Aviation Space Environmental Medicine*, **71** (3), 216–224.

Barish, R. (1996) *The Invisible Passenger: Radiation Risks for People Who Fly*, Advanced Medical Publishing.

Barish, R.J. (1999) In-flight radiation – counseling patients about risk. *Journal of the American Board of Family Practice*, **12**, 195–199.

Blettner, M., Zeeb, H., Auvinen, A. *et al.* (2003a) Mortality from cancer and other causes among male airline cockpit crew. *International Journal of Cancer*, **106** (6), 946–953.

Blettner, M., Zeeb, H., Auvinen, A. *et al.* (2003b). Mortality from cancer and other causes among male airline cabin attendants in Europe: a collaborative cohort study in eight countries. *American Journal of Epidemiology*, **158** (1) 35–46.

Bryson, B. (2003) *A Short History of Nearly Everything*, Doubleday.

Chen, J. and Mares, V. (2008) Estimate of the dose to the fetus during commercial flights. *Health Physics*, **95** (4), 407–412.

Chodick, G., Bekiroglu, N., Hauptmann, M. *et al.* (2008) Risk of cataract after exposure to low doses of ionizing radiation: a 20-year prospective cohort study among US radiologic technologists. *American Journal of Epidemiology*, **168**, 620–631.

Clarke, M. (2003) National Radiological Protection Board e-Journal, No. 4.

Cucinotta, F.A., Manuel, F.K., Jones, J. *et al.* (2001) Space radiation and cataracts in astronauts. *Radiation Research*, **156** (Pt 2), 460–466.

Davis, R.L. and Mostofi, F.K. (1993) Cluster of testicular cancer in police officers exposed to hand-held radar. *American Journal of Industrial Medicine*, **24** (2), 231–233.

Geeze, D.S. (1998) Pregnancy and in-flight cosmic radiation. *Aviation, Space and Environmental Medicine*, **69**, 1061–1064.

Gundestrup, M. and Storm, H.H. (1999) Radiation-induced acute myeloid leukaemia and other cancers in commercial jet cockpit crew: a population-based cohort study. *The Lancet*, **354**, 2029.

Gundestrup, M., Andersen, M.K., Sveinbjornsdottir, E. *et al.* (2000) Cytogenetics of myelodysplasia and acute myeloid leukaemia in aircrew and people treated with radiotherapy. *The Lancet*, **356**, 2158.

Haldorsen, T., Reitan, J.B. and Tuelen, U. (2000) Cancer incidence among Norwegian airline pilots. *Scandinavian Journal of Work, Environment and Health*, **26**, 106–111.

Hammar, N. *et al.* (2002) Cancer incidence in airline and military pilots in Sweden 1991–1996. *Aviation, Space and Environmental Medicine*, **73** (1), 2–7.

Hunter, R. (2003) *Protection of Aircrew from Cosmic Radiation: Guidance Material*, CAA.

Iles, R.H.H.A., Jones, J.B.L., Bentley, R.D. *et al.* (2003) The effects of solar particle events at aircraft altitudes. Proceedings of ESA Space Weather Workshop: Looking towards a European space weather programme (17–19 December 2001), pp. 121–124.

Irvine, D. and Davies, D.M. (1999) British Airways flight-deck mortality study. *Aviation, Space and Environmental Medicine*, **70** (6), 548–555.

Flight Health. The problem of cosmic radiation. flight health.org/cosmic-radiation/cosmic-radiation-the-problem (accessed February 2010).

Lean, G. (2009) Mobiles and cancer: the plot thickens. *Daily Telegraph*, 12 September.

Lim, M.K. and Bagshaw, M. (2002) Cosmic rays: are air crew at risk? *Occupational and Environmental Medicine*, **59**, 428–432.

Linnersjö, A., Hammar, N., Dammström, B.-G. *et al.* (2003). Cancer incidence in airline cabin crew: experience from Sweden. *Occupational and Environmental Medicine*, **60**, 810–814.

Moir, I. and Seabridge A. (2006) *Military Avionics Systems*, John Wiley & Sons, Ltd.

Tokumaru, O., Haruki, K. , Bacal, K. *et al.* (2006) Incidence of cancer among female flight attendants: a meta-analysis. *Journal of Travel Medicine*, **13** (3), 127–132.

Parry, D. (2009) *Moon Shot: The Inside Story of Mankind's Greatest Adventure*, Ebury Press.

Pukkala, E. *et al.* (2003) Cancer incidence among 10,211 airline pilots: Nordic study. *Aviation, Space and Environmental Medicine*, **74**, 699–706.

Rafnsson, V., Olafsdottir, E., Hrafnkelsson, J. *et al.* (2005) Cosmic radiation increases the risk of nuclear cataract in airline pilots. A population-based case-control study. *Archives of Ophthalmology*, **123**, 1102–1105.

Rafnsson, V., Sulem, P., Tulinius, H. and Hrafnkelsson, J. (2003a) Breast cancer risk in airline cabin attendants: a nested case-control study in Iceland. *Occupational and Environmental Medicine*, **60**, 807–809.

Rafnsson, V., Hrafnkelsson, J. and Tulinius H. (2000) Incidence of cancer among commercial airline pilots. *Occupational and Environmental Medicine*, **57**, 175–179

Rafnsson, V., Hrafnkelsson, J., Tulinius, H. *et al.* (2003b) Risk factors for cutaneous malignant melanoma among aircrews and a random sample of the population. *Occupational and Environmental Medicine*, **60**, 815–820.

Sigurdson, A.J. and Ron, E. (2004) Cosmic radiation exposure and cancer risk among flight crew. *Cancer Investigation*, **22** (5), 743–761.

Stroud, R. (2009) *A Book of the Moon*, Doubleday.

Taylor, G.C., Bentley, R.D., Conroy, T.J. *et al.* (2002) The evaluation and use of a portable TEPC systems for measuring in-flight exposure to cosmic radiation. *Radiation Protection Dosimetry*, **99** (1–4), 435–438.

Whelan, E.A. (2003) Cancer incidence in airline cabin crew. *Occupational Environmental Medicine*, **10** (11), 805.

Yong, L.C., Sigurdson, A.J., Ward, E.M. *et al.* (2009) Increased frequency of chromo-some translocations in airline pilots with long-term flying experience. *Occupational Environmental Medicine*, **66**, 56–62.

Further Reading

Fortescue, P., Stark, J. and Swinerd, G. (2003) *Spacecraft Systems Engineering*, John Wiley & Sons, Ltd.
Freidburg, W., Copeland, K., Duke, F.E. *et al.* (2000) Radiation exposure during air travel: guidance provided by the Federal Aviation Administration for air carrier crews. *Health Physics*, **79**, 591–595.
Hunter, R.M.C. (2002) Cosmic radiation, in *Aviation Medicine and the Airline Passenger* (eds A. Cummin and A. Nicholson), Arnold.
May, Captain Joyce. (2005) Considerations Regarding Flying and Pregnancy. *Science Daily*, 1 September.

Useful Web Sites

Aircrewhealth.com
britishairways.com/travel/healthcosmic/public/en_gb
dft.gov.uk
hps.org (Health Physics Society)
science.nasa.gov
solarstormwatch.com
stfc.ac.uk (Science Technologies & Facilities Council)
virgingalactic.com
wddty.com (What doctors don't tell you)
who.int/ionizing_radiation/env/cosmic/WHO_Info_Sheet_Cosmic_Radiation.pdf
www.hpa.org.uk (Health Protection Agency).
www.who.org

7

Back and Neck Pain

7.1 Back Pain

Back pain is a big enough subject to warrant its own book, not just a chapter. In fact, back pain impacts on the global economy and is one of the most commonly cited reasons for absenteeism across the developed world.

7.1.1 Lower Back Pain

Lower back pain is one of the conditions most commonly seen by doctors and because its cause is often unclear, it is most often treated with long-term pain killers or exercise regimes rather than surgery. It affects both men and women and most adult age groups.

In aviation, it is a problem across all communities and can affect not only civil and military aircrew, but even the most infrequent commercial passenger. A recent survey by the British Chiropractic Association (March 2009) shows that 72% of people in the UK say they have suffered from back pain at some point, with more than a third currently suffering. The survey also found that a third of people spend more than 15 hours a day sitting down. British Chiropractic Association members have 60 000 to 70 000 patient appointments between them each week.

Most lower back pain is caused by the downward 'load' on the lower portion of the spine and can often be aggravated by poor posture and by sitting still for long periods.

Interestingly, though, when NASA commissioned a study into back pain in crew members during space station missions, it found that lower back

Air Travel and Health: A Systems Perspective   Allan Seabridge and Shirley Morgan
© 2010 John Wiley & Sons, Ltd

pain was an unexpected side effect of zero gravity. Theoretically, where the gravitational load on the spine is reduced, problems with lower back pain should be alleviated, not aggravated. The study (Snijders, 2010) looked at the possibility of pain being caused by problems with the muscles of the lower back, rather than the actual vertebrae. Muscles lose mass during spaceflight – it could be that the 'girdle' of muscle protecting the lower back is weakened.

It is this muscle pain, rather than problems with the actual skeleton, that is most often the cause of back pain and those with a pre-existing or underlying medical problem are most likely to suffer discomfort caused by flying.

7.1.2 Posture

Posture and sitting position can play a huge part in aggravating muscle pain in the back and also the neck. Sitting in one fixed position and being unable to flex the back and neck for long periods causes aching muscles and stiffness. Draughts on aircraft can be a major exacerbating factor for those prone to neck and backache. Seats deteriorate with use and collapsing seat cushions undoubtedly contribute to discomfort and poor posture.

There have been dozens of studies into flying-induced back and neck ache and its causes, but in all relevant studies of back and neck problems, the measure of pain is a very subjective phenomenon and its true 'severity' is difficult to compare from one person to another.

Figure 7.1 shows the critical element of the neck and back, damage to which can lead to back pain.

The back is a complex structure consisting of:

- Twenty-four small bones (vertebrae) that support the weight of the upper body and form a protective canal for the spinal cord.
- Shock-absorbing discs that cushion the bones and allow the spine to bend.
- Ligaments that hold the vertebrae and discs together.
- Tendons to connect muscles to vertebrae.
- A spinal cord, which carries nerve signals from the brain.
- Nerves and muscles.

7.1.3 Back Pain and Military Aircraft

Back injury can arise from sitting in cramped conditions for many hours strapped into a seat, especially for the military fast-jet pilot strapped tightly into an ejection seat and flying at low level in turbulent conditions. The seat is designed to save life on ejection, not as a comfortable resting place, although efforts have been made to make the seat acceptable. Complaints of chronic

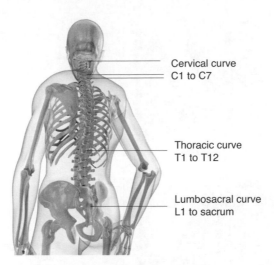

Cervical curve
C1 to C7

Thoracic curve
T1 to T12

Lumbosacral curve
L1 to sacrum

Figure 7.1 Physiology of neck and back (Sebastian Kulitzki). Reproduced with permission from Shutterstock Images. See Plate 13 for the colour figure.

back pain have been reported and there is evidence that this is a factor in early retirements or reduced flying. Complaints have been made by qualified flying instructors (QFIs) because they spend many more hours than their students strapped in their seats.

Back injury arises from long periods of sitting in a confined position during normal operations – the seat lumbar support is not optimised for comfort. The seat is adjustable for height in order to obtain the correct eye-line, but there is no adjustment for the backrest or the position of controls.

Back injury is also a potential problem for surveillance aircraft crew because of the duration of the mission and the need to remain seated at a workstation. The levels of ride quality at low level in high sea states for maritime patrol aircraft such as the Nimrod also cause concern. Some members of the crew of such aircraft types are seated at a right angle to forward motion, often facing outboard.

Back pain associated with military flying has been much studied in recent years and back and neck problems caused by sitting at controls for long periods and also by vibration on the flight deck have been well documented.

7.1.4 Helicopter Pilots

Figure 7.2 shows some of the abrupt changes of attitude to which helicopter pilots and passengers are subjected. Helicopter pilots fare particularly badly when it comes to the incidence of back pain.

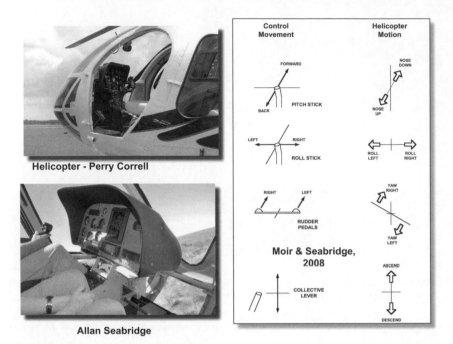

Figure 7.2 Helicopter crew seating and vehicle movements. Reproduced with permission from 1. Allan Seabridge. 2. Courtesy of Shutterstock Images. 3. Moir & Seabridge.

A report into back pain in the naval helicopter sector (Sargent and Bachmann) concluded that long hours in the cockpit, ineffective seat padding, poor posture, use of night vision goggles (NVGs), and constant vibration all may contribute to strain and fatigue in the lumbar region.

It showed that subsequent pain could range from 'mild intermittent annoyance' to something so debilitating that it impacts on flight safety, causing difficulties for squadron operations as a result of hurried or cancelled flights. The study considered that the prevalence of back pain in an otherwise healthy, young population of aviators could be as high as 82 to 92%.

The most common impacts of back pain on flight operations are decreased concentration, hurried flight and, more rarely, cancelled flight. In addition, the Sargent–Bachmann report said that up to 12% of pilots reported sometimes missing work altogether because of the pain.

Interviews were conducted with two female helicopter pilots to gain an insight into the causes of discomfort. Pilot A has over 20 years' experience flying helicopters, primarily as an instructor. She said:

As an instructor you get used to sitting in uncomfortable positions – trainee pilots can struggle to keep the aircraft in balance, so you often find your-self at an angle, tensing your muscles to compensate for the imbalance of the aircraft. After a time it can cause muscle pain, particularly in the lower back.

Adjustable seat support would be good, but weight, safety and cost all have to take priority and any permanent adaptations undertaken by a pilot or operator need approval from the CAA. No operator can afford to have a helicopter grounded while waiting for the approval to fit something like a new lumbar support.

I am sure that there are many pilots who have health problems at-tributable to flying helicopters, but they choose to put up with them. Per-sonal fitness is important, because without your medical clearance you don't have a job; and most of us would never want to lose the view we get from our office window.

Pilot B is a helicopter pilot for the police in the North West. She said:

Flying a helicopter means using tail rotor pedals and the distance between seat and pedals tends to be designed for someone around five feet ten. I'm only five feet four, and in some helicopter types there is no physical adjustment available for the seat or pedals.

The shorter pilot may compensate by sitting in a different position – in some aspects of flying you can become round-shouldered in the seat after a few hours and it puts a strain on the back. The seat starts to feel like a church pew.

Lumbar support and some better ergonomics in the cockpit would help, but helicopter pilots come in a wide range of shapes and sizes, making it hard to find a cost-effective and safe solution for everyone.

7.1.5 Posture and Pain

Posture varies with aircraft and sortie type. There are data to support the idea that back pain varies according to flight regimen, and in other industry sectors it has been suggested that a forward-flexed posture is often associated with chronic pain. This is borne out by studies which suggest that pilots report back pain more often during instrument flying (instrument flight rules – IFR) than during visual flight (visual flight rules – VFR) and looking out of the window.

The cockpit of a fast-jet aircraft is usually constrained by the volume al-located to it in the airframe budget and the shape is determined by the aerodynamic considerations of the fuselage. Located at the 'sharp end' of the aircraft there is very little volume available, and what there is must accommodate the pilot, the ejection seat, the displays and control panels and some key avionic equipment. The seat is located so that the pilot is able to reach the controls and see the display surfaces; this must apply to a percentile range of pilots for which the aircraft has been designed. The recline angle of the seat back is usually set at about 21° to the vertical. This is a good angle for safe ejection lines whilst maintaining a good pilot view in the cockpit. A lower recline angle may help with g-tolerance, but means that the pilot's knees obscure the view of the instrument panels.

The paper 'Task and postural factors related to back pain in helicopter pilots' (Bridger *et al.*, 2002) builds on a previous survey which revealed a high prevalence of back pain in Royal Navy helicopter aircrew. A second survey took account of flying tasks and cockpit ergonomics, with a questionnaire focusing on pain in both the flying pilot and co-pilot/instructor roles. The prevalence of back pain was 80% and task-related back pain was greatest in instrument flying (72%) and least in the co-pilot and instructor roles (24%). Self-ratings of posture indicated that forward-flexed trunk postures predominated in the flying roles and were most extreme in instrument flying. The paper concluded that much of the back pain experienced by helicopter pilots is due to the posture needed to operate the cyclic and collective controls. In instrument flying, it could be that the visual demands of scanning the displays may worsen the pain by causing the pilot to lean further forward.

A study of Indian helicopter pilots (and the subsequent trial of lumbar support cushions) concluded that backache was a 'serious malady' for helicopter pilots, both military and commercial, and was primarily caused by posture-related muscle strain. It recommended a regular regime of exercise and the provision of lumbar support during flight (Sharma and Upadhyay, 2000).

There have been calls for further studies following a comparison of the effectiveness of individually moulded lumbar support in relieving postural backache in RAF aircrew employed in different aircraft types (Graham-Cumming). This study questioned all UK military aircrew issued with lumbar supports for postural backache between 1 January 1986 and 31 January 1995. There were 309 responses (out of 329 questionnaires issued) and these indicated that the support provided relief for between 62.9 and 91.7% of the groups of aircrew studied. There were, however, significant differences between ejection seat aircrew and helicopter pilots and the results suggested that the lumbar supports impaired performance and were unsatisfactory in mobile rear crew. The author recommended further study into back pain suffered by helicopter pilots.

A study into aircraft type and diagnosed back disorders in US Navy pilots and aircrew (Simon-Arndt, Yuan and Hourani, 1997) focused on the extent to which the type of aircraft flown is associated with diagnosed back problems, and examined differences in the prevalence of back disorders between pilots and aircrew. Navy pilots and aircrew members with a diagnosed back disorder on their most recent physical examination between 1991 and 1993 were compared with pilots and aircrew without such diagnoses. Results showed that aircrew had a higher risk of diagnosed back problems than pilots for both helicopters and fixed-wing aircraft – and flight engineers had a higher risk of diagnosed back problems than other aircrew members. Amongst pilots, no association was found between the type of aircraft and diagnosed back problems.

In a separate study, 185 E-2C Hawkeye aircrew volunteered to complete a neck and back pain and symptoms study: 78% of pilots and 74% of naval flight officers (NFOs) reported neck and/or back pain in the previous year. Similarly, 68% of pilots and 70% of NFOs reported having neck and/or back pain symptoms in their past 30 flights. The most common symptom was an in-flight dull ache, lasting for one or two days (Loomis *et al.*, 1999).

One former Tornado navigator with over 2000 flying hours says his chronic lower back pain is partly caused by flying fast jets, partly by physiology:

I'm just under 6'2", but with a relatively long back. I have also been diagnosed with a smaller than average 'canal' which encloses my spinal cord. Add to that, I played rugby for years. Together, those factors have combined to damage my lower back to the extent that my specialist says his aim is to help me avoid surgery before I reach 60.

I do know of people judged to be too tall for fast jets, but when I joined the RAF the measurements taken were really confined to overall leg length – I had to be able to fit the smallest escape envelope, which at the time was the back seat of the Buccaneer. My measurements for that were ok, but having a long back has, I think, affected my posture, making me constantly compensate and impacting on other areas, such as my neck.

In the back seat of a Tornado you spend a lot of time hunched over instruments and as an instructor you tend to sit as high as you can so that you can see what's going on.

Would I have such a bad back if I had never flown? I think it's unlikely, and I know many others with similar problems, but flying fast jets is very competitive and often your success is judged by your peers. You tend to just put up with any discomfort because the military mind is focused on achieving the mission. I was in the RAF for 16 years but never really mentioned it because I just wanted to keep flying and playing contact sports.

It wasn't until later, when I found out more about my own physiology and saw the bigger picture, that I realised how each separate factor contributed and how much damage had been done.

(Author interview)

7.2 Neck Strain

A fighter pilot places the same stresses on his neck as a Formula One racing driver. Any fast-jet pilot needs good flexibility and reaction times, but neck injuries are common, mainly muscle strains brought on during high-*g* turns, braking and accelerating.

Injury to the neck and upper vertebrae may be caused by the effects of high-*g* manoeuvring, especially with a helmet, this situation being exacerbated by additions to the helmet such as sights, night vision devices and displays.

A human head weighs around 6 kg. Add to that a helmet with a helmet-mounted sight or display (HMD) as illustrated in Figure 7.3 and the impact

Eurofighter Typhoon Integrated Helmet-Mounted Display

The Joint Strike Fighter HMD

Figure 7.3 Pilot wearing a helmet-mounted display. Reproduced with permission from Aircraft Display Systems, Jukes 2004.

of high positive g-force and the head can weigh more than 40 kg, putting the upper back and neck under immense strain.

Typical mass figures for this system obtained from Jukes (2004) are:

Human head	10 lb	4.5 kg
Helmet	2 lb	0.9 kg
Display module	1.5 lb	0.7 kg
Oxygen mask	0.6 lb	0.3 kg
Cabling and connectors	1 lb	0.5 kg

There have been examples of extreme damage resulting from a combination of reclined posture, turning of the head during a manoeuvre with the added mass of a helmet, and using a force stick without artificial feel, resulting in rapid onset of g, grey-out, neck muscle relaxation, forced rotation of the head mass against the cervical vertebrae. All of this resulted in a broken neck.

Neck pain is commonly reported by fast-jet pilots. In one study 50% of pilots reported in-flight or immediate post-flight pain and 90% described at least one pain event during high-g turns (Jones *et al.*, 2000).

The pain may be accompanied by headaches but, like lower back pain, is often non-specific. Most is attributed to soft-tissue injury but at least one study (Petren-Mallmin and Linder, 2001) has shown that experienced fast-jet pilots may be more prone to degenerative conditions of the neck than a control group in the general population. The condition 'military neck' is frequently described by chiropractors and refers to loss of curvature and flexibility in the neck muscles, which may be degenerative.

An example of active flying gives a good example of the issues:

The Mk3 helmet (which I wore for about 20 years on and off) weighed nearly 7lbs. This meant in air combat that your head plus helmet weighed some 150lb under the force of 7g. this was not so bad if you had the luxury of being able to sit straight and keep your head permanently still until the g came off. I think that some of our aviation medics really imagined this was what we did in combat. Fat chance. In air combat you were constantly twisting and turning your head around almost through 180° under g to see what was going on and all the while with this huge load on your neck. Time after time we told the doctors that this was literally a pain in the neck and we wanted the American style lightweight helmets.

(Pook, 2009)

Just as exercise regimes have been found to help lower back pain, specialist weight-training equipment designed to strengthen the upper back and neck has been trialled with the RAF.

The Royal Norwegian Air Force has attempted to tackle this issue by recommending specific physical training patterns to reduce the incidence of neck pain (Hansen and Wagstaff, 2001).

Head movements in the cockpit, combined with high-g manoeuvres, are associated with neck pain in the fighter pilot community. Research has shown that neck injury incidents have increased as Hawk pilots are introduced to Typhoon training. A treatment protocol has been proposed which will reduce the recovery time and help to reduce the incidence of recurrence. However, this treatment does not prevent the occurrence of the problem. There is evidence that the RAF is also looking at exercise as a measure to reduce occurrence (Morris, 2010).

F-16 pilots have a high incidence of minor neck injuries. It was hypothesised that pilots who did neck-strengthening exercises and pilots who used other preventive strategies would have fewer injuries. In a survey of 268 US Air Force F-16 pilots, subjects were divided into two groups. Group I, the Early Intervention Group, performed an intervention, or not, from the start of their F-16 careers. Outcomes were measured as a percentage of pilots reporting an injury during their F-16 careers. Group II, the Midstream Intervention Group, initiated an intervention after sustaining an injury. Injuries before and after the intervention were compared as a median injury rate per 100 hours F-16 time. Results showed that the one-year prevalence of neck injury was 56.6% and for an F-16 career was 85.4%. For every 100 hours in the F-16, the risk of injury increased by 6.9%. Only 26.9% of the pilots routinely did neck-strengthening exercises. For the Early Intervention Group, fewer injuries were associated with neck-strengthening exercises and placing the head against the seat prior to loading +Gz. For the Midstream Intervention Group, a lower median injury rate was associated with neck-strengthening exercises, placing the head against the seat prior to loading, warming up with stretching or isometrics, pre-positioning the head prior to loading, and unloading prior to moving the head. Interventions not associated with fewer injuries included body exercises and placing the head against the canopy. The conclusion reached was that certain strategies may prevent neck injuries. More research is needed to confirm these results (Albano and Stanford, 1998).

7.3 Commercial Aircraft Issues

7.3.1 Flight Attendants

If most back problems are caused by sitting for long periods in cramped conditions or by being exposed to vibration and high-g manoeuvres, flight attendants might be expected to be immune from the worst effects.

However, a study into the health of Sri Lankan flight attendants in 2007–2008 showed that many suffered from back problems. Many of the problems were caused by back injury associated with their duties – pushing the service trolley or lifting heavy items, for instance – but over half of those questioned said they suffered repeated backache, ranging from 'mild' to 'severe' (Agampodi, Dharmaratne and Agampodi, 2009).

7.3.2 Passengers

In a survey conducted by SpineUniverse in the summer of 2008, 88% of people who had flown in North America in the previous year said they had back or neck pain, or both, after a typical flight. The spine-universe survey (spineuniverse.com/conditions/back/pain/effect/flying-back-neck-pain-survey) asked respondents if they would be willing to pay more to be guaranteed a seat with added comfort features for back and neck pain sufferers. Not surprisingly, 74% said yes – but only wanted around $50 more added to their ticket price.

People who suffer from lower back or neck pain can find the cramped conditions of an aircraft cabin uncomfortable and almost unbearable in some cases. Online review sites and blogs are full of complaints from paying passengers who claim their flying experience – and in some cases their holiday – was ruined by backache. Posture is important in avoiding back pain, but airline seats outside of first or business class do not have the range of movement or built-in support required to accommodate everyone in comfort. A seat with a 28 inch pitch gives little opportunity to stretch the spine and most seats are produced around a one-size-fits-all premise.

Figure 7.4 shows some examples of seats in different classes.

7.4 Lumbar Support

Could the problem be solved by adjustable lumbar support and a movable headrest – the same sort of seating flexibility you get in the driver's seat of a modern car? Lumbar cushions have been proved to have a positive impact on military helicopter pilot comfort, but could something similar be made available on passenger flights?

There are examples of lumbar support systems in the marketplace. Though the focus has inevitably been on the comfort of first- and business-class passengers, where lie-flat beds are now widely available, products such as FutureFlite are being promoted to economy airlines. The system does not require any electric power and can provide a variety of tilt positions and adjustable

Standard class seats - Magicinfoto

**Business class seats
– Vladimir Sazonov**

Business class seats – Petronito G. Dangoy Jr

Figure 7.4 Various classes of airline seating. Reproduced with permission from Shutterstock Images. See Plate 9 for the colour figure.

support for all shapes and sizes. The system consists of a simple assembly with minimum components using a spring-loaded action and remote push-button control compatible with industry-standard recline cables. The system, shown in Figure 7.5, offers the following features:

- Smooth, fast-acting and responsive device to permit frequent and easy adjustment using the pushbutton, with the passenger leaning back to counteract the spring.
- A fore and aft tilting motion to accommodate the spinal curvatures of all passengers which eliminates the need for vertical adjustment.
- Infinite adjustment of lumbar support.
- Cushioning effect of shock-absorbing 'microlock' which makes the lumbar support comfortable even with worn and fatigued cushions.

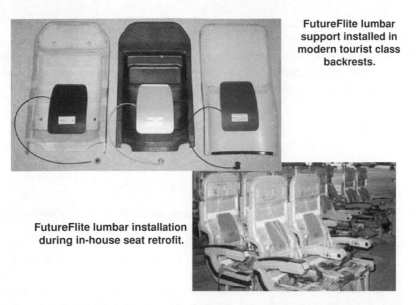

FutureFlite lumbar support installed in modern tourist class backrests.

FutureFlite lumbar installation during in-house seat retrofit.

Figure 7.5 Installing lumbar support. Reproduced with permission from FutureFlite Corp.

7.5 Advice for Passengers

How do passengers avoid back and neck ache during a flight? As with conditions such as DVT, they are often prevented from moving around the cabin during flight because of security and safety restrictions or the presence of the drinks trolley.

The most obvious solution is to stand up, walk and stretch as much as you can. Those with serious back problems could consider travelling with a doctor's note to alert the flight attendants of the need to move around the cabin. In that case, an aisle seat would be a wise choice.

Passengers can ask for extra pillows to put behind the back to keep the spine straight, as this will help to ease the pressure on it. It is also good practice to keep the legs at a right angle when sitting, using a bag or pillow to prop up the feet if necessary (though you may be asked to keep the floor space of the row clear). Anyone who is tall should try and get an exit row or bulkhead seat with more legroom.

Some people recommend using hot and cold packs during the flight to heat and cool the muscles of the back and neck (but it would be wise also to pack a doctor's note for airport security; they would not take kindly to gel packs in hand luggage). Inflatable pillows are widely available at airport and travel shops and claim to relieve neck stresses.

Protection from cold draughts is also essential. Take a scarf or blanket but remember that although they offer more space to stretch, seats near emergency exit doors are likely to be colder.

References

Agampodi, S.B., Dharmaratne, S.D. and Agampodi, T.C. (2009) Incidence and predictors of on-board injuries among Sri Lankan flight attendants. *BMC Public Health*, **9**, 227.
Albano, J.J. and Stanford, J.B. (1998) Prevention of minor neck injuries in F-16 pilots. *Aviation, Space and Environmental Medicine*, **69**, 1193–1199.
Bridger, R.S., Groom, M.R., Jones, H. *et al.* (2002) Task and postural factors are related to back pain in helicopter pilots. *Aviation, Space and Environmental Medicine*, **73** (8), 805-811.
Graham-Cumming, A.N. Moulded lumbar supports for aircrew backache: comparison of effectiveness in fixed and rotary wing aircrew, Headquarters, Personnel and Training Command, RAF Innsworth, Gloucester. http://ftp.rta.nato.int/public/PubFullText/RTO/MP/RTO-MP-019/$MP-019-35.PDF (accessed May 2010).
Hansen, O.B. and Wagstaff, A.S. (2001) Low back pain in Norwegian helicopter aircrew. *Aviation, Space and Environmental Medicine*, **72**, 161–164.
Jones, J.A., Hart, S.F., Baskin, D.S., Effenhauser, R., Johnson, S.L., Novas, M.A., Jennings, R. and Davis, J. (2000) Human and behavioral factors contributing to spine-based neurological cockpit injuries in pilots of high-performance aircraft: recommendations for management and prevention. *Military Medicine*, **165** (1), 6–12.
Jukes, M.L. (2004) *Aircraft Display Systems*, John Wiley & Sons, Ltd.
Loomis, T.A., Hodgson, J.A., Hervig, L. and Prusacyck, W.K. (1999) *Neck and Back Pain in E2-C Hawkeye Aircrew*, Storming Media.
Moir, I. and Seabridge, A. (2008) *Aircraft Systems*, 3rd edn, John Wiley & Sons, Ltd.
Petren-Mallmin, M. and Linder, J. (2001) Cervical spine degeneration in fighter pilots and controls: a 5-yr follow-up study. *Aviation, Space and Environmental Medicine*, **72** (5), 443–446.
Pook, J. (2009) *Flying Freestyle: An RAF Fast Jet Pilot's Story*, Pen and Sword Books.
Morris, C.B., Wg Cdr (2010) RAF Aviation Medicine Training of RAF Aircrew. ftp.rta.nato.int/public/pubfulltext/rto/mp/rto-mp-021 (accessed May 2010).
Sargent, P., Lt MD and Bachmann, A., Lt MD (n.d.) Back pain in the naval rotary wing community. Naval Safety Center, http://safetycenter.navy.mil/Aviation/articles/back_pain.htm (accessed May 2010).
Sharma, S., Wg Cmdr and Upadhyay, A.D., Wg Cmdr (2000) Is backache a serious malady among Indian helicopter pilots and low backache among Chetak helicopter

pilots? Trial of lumbar cushions at a flying unit. *Indian Journal of Aerospace Medicine*, **44**, 56–63.

Simon-Arndt, C.M., Yuan, H. and Hourani, L.L. (1997) Aircraft type diagnosed back disorders in US Navy pilots and aircrew. *Aviation, Space and Environmental Medicine*, **68**, 1012–1018.

Snijders, C.J. (2010) Study of low back pain in crew members in space flight. International Space Station fact sheet, http://www.nasa.gov/mission_pages/station/science/experiments/Mus.html (accessed May 2010).

Further Reading

Davies, J.R., Johnson, R., Stepanek, J. and Fogart, J.A. (2008) *Fundamentals of Aerospace Medicine*, 4th edn, Walters Kluwer.

Nachensson, A. and Jonsson, E. (2000) *Neck and Back Pain: The Scientific Evidence of Causes, Diagnosis and Treatment*, Lipincott, Williams and Wilkins.

Scott, W.B. (2004) Pulling gs on earth. *Aviation Week and Space Technology*, 5 January.

8

Military Issues

The military fast jet imposes some tough conditions on the cockpit inhabitants – so tough in the early days of supersonic flying in the UK that 'The country's leading aviation establishment had already predicted that pilots would be unable to withstand the effects of flying aircraft at such high speeds' (Barnett-Jones, 2008). This thinking led to the cancellation of the Miles M52 supersonic project, although the English Electric Lightning survived. Since then many supersonic aircraft have been designed and are operated on a daily basis, including some types that fly very high and fast to perform specialised surveillance missions and to avoid detection by enemy radar and missile defence systems. The space programme has shown that environments can be designed to support human life for long periods of time.

Nonetheless, there are some specific conditions that are imposed on military fast-jet pilots that differ considerably from the commercial airliner environment. The fast-jet pilot is subject to high positive and negative acceleration forces, high temperatures and rapid changes of altitude and temperature. The pilot also can spend a long time firmly secured in a small cockpit without any of the trappings of comfort such as food, entertainment and freedom to move around. Some of the effects of this environment have been covered in previous chapters: air quality in Chapter 3, noise and vibration in Chapter 5, radiation effects in Chapter 6, neck and back pain in Chapter 7. These conditions will not be repeated here, whilst conditions specific to fast-jet crews are included below.

Air Travel and Health: A Systems Perspective   Allan Seabridge and Shirley Morgan
© 2010 John Wiley & Sons, Ltd

It must be emphasised that aircrew are selected for fast-jet service if they are demonstrably fit. They must continue to train and maintain fitness throughout their service career, and they will be subject to regular medicals.

8.1 The Cockpit Environment

The cockpit of the modern fast jet is cramped, noisy and hot, whilst the aircraft performance can impose rapid and sustained changes of attitude and acceleration on the pilot. In addition the aircraft can undergo rapid changes of altitude and cockpit pressure many times in a mission. There is a need to fly the aircraft, to concentrate on the mission and to assimilate information and make rapid decisions throughout. These decisions not only affect the success of the mission, but also affect flight safety and hence the safety of the inhabitants of the aircraft and the over-flown population.

The aircraft, the cockpit and the clothing worn by the pilot have evolved to protect the pilot from the rigours of supersonic flight. Test pilot Jimmy Dell described kitting up for a Lightning development flight:

> Because the tests are going to take me out over water and the sea temperature is below 50°F, I shall need to wear special equipment. This consists of an anti-exposure suit over my air ventilated garment (which bleeds conditioned air to the body), my flying suit, anti-g trousers and, over the trousers, leg restraint garters and an important garment, my Mae West life jacket. This is equipped with a personal locator beacon and mini-flares. The oxygen tube is attached to the life jacket and constitutes what is called the 'man portion' of the personal equipment connector (PEC). When plugged into the 'aircraft portion' of the PEC this provides oxygen, radio communication and air conditioning. On the way out I pick up my helmet . . .
>
> (Barnett-Jones, 2008)

This combination of clothing and equipment is known as the aircrew equipment assembly (AEA).

The cockpit of a military fast-jet aircraft subjects the crew to high heat loads. Figure 8.1 gives some indication of the external and internal sources of heat including exposure to sunlight through the canopy, hot conditions during taxiing after starting up in hot ground conditions, dissipation from equipment and conditioning system extremes.

Aircrew can suffer from heat stress as a result of a combination of their clothing (e.g. flying suit, inflatable garments, immersion suit, winter

❑ Solar input

❑ Kinetic heating

❑ Equipment dissipation

❑ Person (metabolic, clothing)

❑ **Total heat load ca 8kW**

❑ To maintain the cockpit at 20-25°C needs
an air flow of 0.25kg/min at 3°C

Figure 8.1 Sources of heat that affect the fast-jet pilot. Reproduced with permission from BAE Systems.

coverall, etc.) and the need to spend long periods of time in an enclosed cockpit in direct solar radiation. This situation is exacerbated on the ground during long-duration pre-flight preparation, taxiing and holding, especially in desert regions where temperatures after a long heat soak can be in excess of 50 °C and where taxiways on large military bases can be many kilometres:

> To imagine the heat stress on a pilot, picture yourself in the heat of summer wearing heavy clothing inside a closed, poorly ventilated plastic bubble, carrying out the most violent physical exercise possible.
>
> *(Pook, 2009)*

This issue is complicated by the fact that a customer tends to select the clothing for operational aircrew, so the aircraft designer does not have control of the total situation.

The pilot's metabolism plays a large part in the heat input; during times of stress such as combat or combat training, the pilot may well hyperventilate and will expend a lot of energy combating g. The overall heat load leads to loss of concentration and stress.

Aircraft designed for high-altitude operations for photographic reconnaissance missions, such as the Lockheed SR-71, pose significant problems for designers. The SR-71 was designed to fly up to 80 000 ft and to achieve speeds of Mach 3.2, reaching the area of Figure 3.2 labelled 'Occasional transits by military aircraft'.

The life support system design for this aircraft was driven by two specific requirements determined by the need to escape safely in the event of an accident:

1. With a standard pressure demand oxygen mask, human lungs cannot absorb enough of 100% oxygen at the operating altitude of this aircraft to sustain consciousness and life.
2. The instant heat rise pulse on the body when exposed to a Mach 3.2 airflow during ejection would be about 230 °C.

As a solution to these issues the crew wore protective full pressure suits throughout the mission. The cabin pressurisation schedule allowed cabin pressure to be selected to an altitude of 10 000 ft or 26 000 ft during flight. This allowed crews flying a low-subsonic flight such as a ferry mission to wear either their full pressure suit or standard AEA.

For this type of aircraft the heating effects are more severe than most fast-jet types. For example, at Mach 3.2 cruise the external heat rise due to kinetic heating creates a surface heat well above 260 °C and a temperature at the inside of the windshield up to 120 °C.

Designers must ensure that the cockpit is designed to provide an environment that protects the pilot from the normal variations in pressure and temperature to be expected in a range of typical missions. The design of the pressurisation systems and seals on the canopy must minimise leakage and prevent catastrophic loss of pressure. The environmental control system must provide a spray of warm or cool air over the pilot with sufficient flow to maintain comfortable conditions. In extreme conditions it may be necessary to design a flying suit that is heated or cooled by the ECS.

More importantly, the system must also take into consideration the total loss of pressure as a result of canopy failure or canopy jettison and must protect the pilot during ejection.

8.2 Effects of Acceleration

Military fast-jet aircrew are subject to very high g, and rapid onset of g, during high-speed manoeuvres. Despite the development of garments to reduce the impact of g, there have been complaints of severe pain in the arms and hands of pilots with the arms stretched out to the stick and throttles.

Although not often experienced in general aviation, military pilots operate at high speeds and undertake manoeuvres that subject them to high g (gravitational) forces. In a vertical climb, the increased g forces (called positive g forces because they push down on the body) tend to force blood out of the circle of Willis supplying arterial blood to the brain. The circle of Willis is an arterial circle at the base of the brain that receives blood pumped up from the two carotid arteries. All the principal arteries that supply the cerebral hemispheres are connected to the circle of Willis (Medicine.net).

The loss of oxygenated blood to the brain eventually causes pilots to lose their field of peripheral vision. Higher forces cause blackouts or temporary periods of unconsciousness. Pilots can use special abdominal exercises and g suits (essentially adjustable air bladders that can constrict the legs and abdomen) to help maintain blood in the upper half of the body when subjected to positive g forces.

In a dive, a pilot experiences increased upward g forces (termed negative g forces) that force blood into the arterial circle of Willis and cerebral tissue. The pilot tends to experience a red-out. Increased arterial pressures in the brain can lead to stroke. Although pilots have the equipment and physical stamina to sustain many positive g forces (routinely as high as five to nine times the normal force of gravity), pilots experience red-out at about 2–3 negative g. For this reason, manoeuvres such as loops, rolls and turns are designed to minimise pilot exposure to negative g forces.

A civilian experiencing the effects of acceleration, both positive and negative g, in a Hawk aircraft vividly describes the experience:

Coming over the top [of a loop] I had what the pilot described afterwards as a grey-out, where the capillaries at the back of the eye are compressed to a point where vision blurs and momentarily dissolves into a grey vortex, and on the way down pulled four gs, which is a sensation like no other, an inescapable, stifling physical presence that pins you to your seat and takes your breath away, as though you are being crushed by some misanthropic phantom … we swooped down to 1,000 ft at 500 knots, then pulled back the stick and shot directly upward, exactly like a rocket, to 17,000 ft. The

manoeuvre took seconds, but felt more like aeons and as the plane gradually slowed I was left with a disorientation so complete that my mind
simply abandoned ship. I was seated, but there was no pressure anywhere
on my body, just a sense of floating in a horizonless, uniform, azure void
with nothing to provide perspective or evidence of where the world was,
or even if it existed outside of myself ... on the way back, having tumbled
out nose first and dropped like a brick, we made 6g, and I blacked out
properly, to wake with no control over my limbs. Pilots call this 'lock-in'
and it typically lasts ten seconds. If it happens to one of them flying solo,
they're unlikely to survive it.

(Smith, 2005)

Pilots are given training to expose them to the effects of acceleration in a
centrifuge, and they can acquire skills to use in flight, which involves a
technique of 'straining' to reduce the volume of blood leaving the brain – the
anti-g straining manoeuvre (AGSM). A description of this manoeuvre can be
found at www.tpub.com/content/aviation2/P-868/P-8680094.htm:

1. A continuous and maximum contraction (if necessary) of all skeletal
muscles including the arms, legs, chest and abdominal muscles (and any
other muscles if possible). Tensing of the skeletal muscles reduces blood in
the g dependent areas of the body and assists in retaining or returning blood
to the thoracic (chest) area, the heart and the brain.

2. The respiratory component of the AGSM is repeated at 2.5 to 3.0 second intervals. The purpose of the respiratory component is to counter the
downward g force by increasing chest pressure by expanding the lungs. This
increased pressure forces blood to and from the heart to the brain.

The respiratory tract is an open breathing system which starts at the nose
and ends deep in the lungs. The respiratory tract can be completely closed
off at several different points. The most effective point is to close the system
off at the glottis. Closing the glottis (which is located behind the 'Adam's
Apple') yields the biggest increase of chest pressure. You can find it and
close it off by saying the word 'Hick'. This should be said following a deep
inspiration and forcefully closing the glottis as you say 'Hick'. Bear down for
2.5 to 3.0 seconds and then rapidly exhale by finishing the word 'Hick'. This
is immediately followed by the next deep inhalation repeating the cycle. The
exhalation and inhalation phase should last no more than 0.5 to 1.0 second.

This is an exhausting technique but some pilots can withstand more than 6g.
Some real-life examples can readily be found on YouTube.

A more controlled mechanism for avoiding the effects of g is the wearing of inflatable garments that constrict the chest and thighs to maintain blood in the region of the upper body. To ensure that the pilot is physiologically compatible with Eurofighter Typhoon's agile capability, the life support system provides pressure breathing and g protection to the extent that 'pilot straining' is not necessary under high-g force. All aircrew equipment is designed as an integral part of the weapon system. It is unique to the Eurofighter Typhoon and includes:

- Full-cover anti-g trousers and chest counter-pressure garment.
- Liquid conditioning garment.
- Nuclear, biological, chemical protection.
- A lightweight head equipment assembly (HEA).

Illustrations of Typhoon aircrew clothing are shown in Figure 8.2.

The system designer must ensure that the appropriate clothing is provided to protect the pilot. If the customer wishes to provide its own AEA, then the designer must make clear what the requirements and limitations are. Crew members must avail themselves of training and maintain a fitness regime.

Geoffrey Lee, Plane focus.

BAE Systems

Figure 8.2 Some examples of aircrew clothing. Reproduced with permission from Geoffrey Lee, Planefocus and BAE Systems. See Plate 14 for the colour figure.

The effects of rapid acceleration and declaration during an ejection are not discussed here, since they have been considered by the authors to be an unusual event, the consequences of which have been well documented elsewhere.

8.3 Pressure Oxygen Breathing and Hypoxia

Fast-jet aircrew are supplied with pure gaseous (liquid oxygen or Lox) oxygen or air with a high concentration of oxygen under pressure that leads to very high dissolved oxygen levels in the blood. On the other hand, under extremes of manoeuvring the brain may be deprived of oxygen. These conditions may occur many times in a flight. The oxygen is mixed with air before being supplied to the pilot and modern aircraft provide oxygen-enriched air from an on-board oxygen generation system (OBOGS), see Chapter 3.

8.3.1 Hypoxia

Breathing at reduced atmospheric pressure results in a reduction of alveolar oxygen pressure, which in turn leads to an oxygen supply deficiency in the body and brain tissues. This condition is termed hypoxia. If the condition persists, then loss of consciousness can occur and, in the extreme, long-term oxygen deprivation leads to brain death.

A more insidious effect of hypoxia is a gradual onset which can lead to performance degradation without the pilot's knowledge. This degradation includes poor visual and physical performance and loss of judgement which may lead to an accident. A famous incident is the loss of a Learjet carrying the golfer Payne Stewart which crashed in October 1999. The aircraft flew for some time after radio contact was lost and a USAF aircraft was scrambled to intercept it. Eventually the aircraft crashed and all occupants were found to have died from suffocation (airsafe.com).

Military pilots undergo a series of exercises in high-altitude-simulating hypobaric (low-pressure) chambers to simulate the early stages of hypoxia (oxygen depletion in the body). The tests provide evidence of the rapid deterioration of motor skills and critical thinking ability when pilots undertake flights above 10 000 ft above sea level without the use of supplemental oxygen. Hypoxia can also lead to hyperventilation as the body attempts to increase breathing rates.

8.3.2 Decompression Sickness

Altitude-induced decompression sickness is another common side effect of high-altitude exposure in unpressurised or inadequately pressurised aircraft. Although the percentage of oxygen in the atmosphere remains about 21% (the other 79% of the atmosphere is composed of nitrogen and a small amount of trace gases), there is a rapid decline in atmospheric pressure with increasing altitude. Essentially, the decline in pressure reflects the decrease in the absolute number of molecules present in any given volume of air.

Decompression sickness in normal operations can occur when an aircraft climbs to very high altitudes where the cabin altitude departs from the nominal 8000 ft to as much as 23 000 ft, and then descends again to a nominal 8000 ft position. A high-performance aircraft zoom climb to 65 000 ft is such an example. This can cause nitrogen to be released into the blood stream, which can cause joint pain. It is similar to, but not as excessive as, the 'bends' suffered from divers ascending too rapidly. This was a fairly common condition is early aircraft which were unpressurised but capable of transitions to 30 000 ft with the pilot breathing a liquid oxygen and air mixture. Early versions of the Jet Provost primary jet trainer are an example.

Rapid changes in altitude allow trapped gases to cause pain in joints in much the same way – although to a far lesser extent – that the bends cause pain in scuba divers. Lowered outside atmospheric pressure creates a strong pressure gradient that permits dissolved nitrogen and other dissolved or trapped gases within the body to attempt to bubble off or leave the blood and tissues in an attempt to move down the concentration gradient towards a region of lower pressure.

Aircrew, cabin crew or passengers participating in any form of diving in their leisure time, especially close to a flight, may find themselves susceptible to dissolved gases in their blood with subsequent risk to health. Many tour operators advise against scuba diving for two days before a flight.

References

Barnett-Jones, F. (2008) *Tarnish 6: The Biography of Test Pilot Jimmy Dell*, Old Forge Publishing.

Pook, J. (2009) *Flying Freestyle: An RAF Fast Jet Pilot's Story*, Pen and Sword Books.

Smith, A. (2005). *Moon Dust*, Bloomsbury.

The Anti-G Straining Manouevre: www.tpub.com/content/aviation2/P-868/P-8680094.htm (accessed 11 May 2010).

Further Reading

Aviation Physiology (2003) World of Earth Science. The Gale Group Inc. Retrieved 8
    December 2009 from Encyclopedia.com: encyclopedia.com/doc/1G2-3437800055.
Vann, R.D., Gerth, W.A., DeNoble, P.J. *et al.* (2004) Experimental trials to assess the risks
    of decompression sickness in flying after diving. *Undersea and Hyperbaric Medicine*,
    31 (4), 431–444.

Useful Web Sites

Airsafe.com
Medicine.net

9

Workstation Use

Have you ever travelled in economy and tried to use your laptop during the flight? It might be allowed, but the cramped conditions, flimsy tray table and poor lighting can make it almost impossible, and certainly most uncomfortable. First- and business-class passengers fare better, but those in economy should probably settle for watching the in-flight movie – though your ability to do that in comfort will depend on where you sit and who you fly with.

However, the difficulties of personal workstation use certainly do not deter everyone. News channels reported in late 2009 that the pilots of a US commercial jet that overshot its destination by 150 miles (240 km) said they were using their laptops and apparently 'lost track of time and their location'.

Their laptop use was of course against regulations, but this is not the only time a portable computer has apparently caused a near accident. In 2008 Australian air safety officials said laptops interfering with the autopilot could have caused a jet suddenly to gain and lose altitude. Dozens of passengers were injured when they slammed into the ceiling.

Potential disasters aside, the everyday use of workstations in flight can cause health problems. The use of workstations and display screen equipment (DSE) in the workplace is covered by health and safety legislation in the UK. This legislation is usually applied to an office environment where environmental conditions are maintained at even temperatures, the office does not move or vibrate and the work area has been designed to provide working conditions that meet health and safety requirements.

Aircrew and mission system operators in aircraft, especially long-duration surveillance aircraft, are expected to spend many hours in their airborne

Air Travel and Health: A Systems Perspective   Allan Seabridge and Shirley Morgan
© 2010 John Wiley & Sons, Ltd

'office' performing high-vigilance tasks. It stands to reason that visual fatigue – like any type of tiredness – must impact on pilot and aircrew performance, but there is evidence that this pattern of work exceeds that permitted by law for office workers.

Commercial passengers will use the in-built entertainment screens and may also be inclined to use their own laptops to do work during the flight. This chapter will examine the conditions under which workstations and laptops are used in flight and will compare this with ground use.

9.1 The Environment

Many military aircraft are now equipped with display screens that are used under conditions that are not optimum for 'office' equipment. Aircraft have a wide range of lighting, wide vibration spectrum, seating optimised for escape rather than comfort, and may be in continuous use for long periods of time. This is especially so in surveillance aircraft such as the Nimrod or E-3 Sentry (Figure 9.1) in which aircrew and mission crew are expected to work for up to

Nimrod MRA4 – Maritime patrol type
BAE Systems

E-3 Sentry – AEW type
U.S. Air Force photo, Senior
Master Sgt. Robert J. Sabonis

E-6B JSTARS

U.S. Air Force photo, Senior
Airman Clark Staehle

Figure 9.1 Examples of surveillance types. Reproduced with permission from 1. BAE Systems. 2. U.S. Airforce photo/Snr. Master Sgt. Robert J Sabonis. 3. U.S. Airforce photo/Snr. Airman Clark Staehle.

10 hours – longer with refuelling. In commercial aircraft the use of computers may be confined to leisure pursuits; even so, a passenger on a long flight may spend several hours watching the screen.

DSE is sometimes referred to as visual display units (VDUs) or computer workstations and includes laptops, touchscreens and other similar devices that incorporate a display screen. Any item of computer-related equipment including the computer, display, keyboard, mouse, desk and chair can be considered part of the DSE workstation. Under UK health and safety legislation, workstations must meet certain basic requirements that enable them to be appropriately adjusted and used without unacceptable risks to health and safety.

A user is defined as an employee who habitually uses DSE as a significant part of their normal work. If someone uses DSE continuously for periods of an hour or more on most days worked, they are likely to be classified as a user.

9.2 Aircraft Environments

9.2.1 Commercial Aircraft

The commercial aircraft cabin has been designed to accommodate passengers in various classes and to seat them comfortably and safely throughout the flight. Seats are predominantly designed to provide upright posture with an option to recline the back through a range of angles determined by the class of cabin. Seat belts and the number of seats in a row restrict movement, as does the seat pitch and the size of adjacent passengers.

In some classes an entertainment screen is provided in the seat back of the passenger in front, or as a fold-away screen in the seat. For standard class the screen is usually located in the ceiling of the cabin.

Passengers, especially in business class, will insist on using their laptops during the journey. The aircraft seats, tables and lighting have not been designed with this purpose in mind, and hence the conditions of use are outside those of an office designed to meet health and safety legislation.

A combination of factors makes the use of laptops with any degree of comfort almost impossible. The table only has a small degree of fore and aft movement relative to the seated position. It will almost certainly not be firm, being designed for accommodating trays of drinks, and it will not provide any support for the wrists. The relative height of the table and seat is fixed and many people will resort to using their laptop on their knees. This results in a hunched position over the keyboard with unsupported arms. Add poor lighting, vibration and occasional turbulence to the mix, and you have a recipe for discomfort.

9.2.2 Military Aircraft

There are military aircraft that perform long-duration reconnaissance tasks which necessitate the use of sensors and on-board databases combined with incoming intelligence messages. This requires a crew of sensor operators and one or more crew members who collate the data, turning the data into information which can be used to support intelligence or operational tactics. Of course, sustained visual monitoring of any kind can cause eye strain – a workstation does not have to be involved. Typical surveillance aircraft roles include:

- Maritime patrol – an aircraft equipped to locate, identify and track maritime assets such as ships, submarines and vessels conducting hostile or illegal operations such as drug or illegal immigrant smuggling, fishing or oil dumping. These types are often used to conduct or co-ordinate search and rescue operations. They are equipped with a range of sensors such as radar, acoustics, electro-optics, electronic support measures and magnetic anomaly detectors. The sensor data are assessed by operators using workstations and co-ordinated to derive a composite picture.
- Airborne early warning (AEW) – an aircraft equipped to locate, identify and track airborne targets that may be a threat in a particular theatre of operations. These types are equipped with a range of sensors such as radar, electro-optics and electronic support measure systems. The sensor data are assessed by operators using workstations and co-ordinated to derive a composite picture.
- Battlefield surveillance – an aircraft equipped to locate, identify and track targets on the ground battlefield that may be a threat in a particular theatre of operations. These types are equipped with a range of sensors such as radar, electro-optics, moving target indicators and signals intelligence systems. The sensor data are assessed by operators using workstations and co-ordinated to derive a composite picture.

All of these types will be equipped with a cabin with a number of installed workstations at which the mission crew do their tasks.

In military applications the users of workstations do not have a choice of whether or not to use the workstation, or for how long, or to have breaks. If the aircraft is operational then the task must be done. This may mean spending

JSTARS
U.S. Air Force photo.
Staff Sgt. Jason Barebo

Nimrod MRA4 BAE Systems

JSTARS
U.S. Air Force photo.
Staff Sgt. Aaron Allmon II

Figure 9.2 Examples of military aircraft workstations in use. Reproduced with permission from 1. BAE Systems. 2. U.S. Air Force photo/Staff Sgt. Jason Barebo. 3. U.S. Air Force photo/Staff Sgt. Aaron Allmon II. See Plate 15 for the colour figure.

up to 18 hours tracking and designating targets in all types of weather conditions. In maritime patrol aircraft this will include extreme manoeuvres and operations at low level in adverse weather conditions. The concentration needed to study the screen and input commands is high and the operator's vigilance needs to be consistent over long periods of time. Extensive studies have been carried out looking at workload issues and human factors when it comes to on-board computer operation. Many focus on how operators can handle multiple demands in terms of incoming data and decision making. Some example workstations are shown in Figure 9.2.

9.3 The System

The task of using a workstation in flight varies according to the type of mission or role that the aircraft is performing. In urban environments, police forces use helicopters for crime prevention and detection and will use a helicopter equipped with electro-optical and night vision sensors. Their task is to track a real-time image on their screen and maintain communications with ground forces. The mission is generally of short duration. Larger aircraft may be used to support operations such as detecting and tracking vessels engaged in illegal activities such as smuggling of illegal immigrants, drugs, contraband and arms, as well as operations concerned with illegal fishing or environmental contamination. These activities may require more sensors and a more complex situation to interpret. Large military surveillance types will be engaged in tracking, detecting and maybe attacking airborne, battlefield or maritime surface or subsurface threats and will deploy many more sensors and use of intelligence to build up a picture of a scenario over many hours or even days. This requires intense concentration over long periods of time to decipher the real-time information, mix it with historical information and build up a picture such as that shown in Figure 9.3.

The latter scenario requires crew members to concentrate on the display, reading the picture and the tabular information, and to deal with communications within the aircraft as well as external communications. Interaction with the displayed information will need some typing of messages and commands,

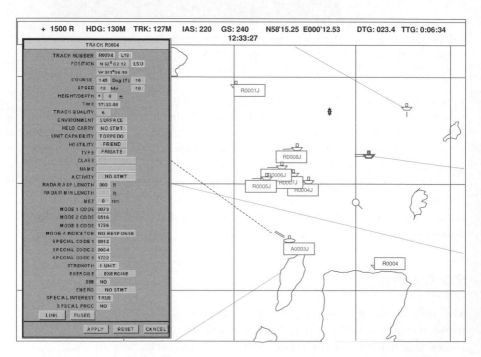

Figure 9.3   Example of a tactical display (after Moir and Seabridge, 2004).

and moving the cursor will usually be performed using a trackball rather than a mouse.

The system of interest is the cabin, especially that part of the fuselage where work is performed by the mission crew, a crew which may number from one person up to more than 10 people. This area is usually equipped with a number of workstations to display the information provided by sensors and at which the crew work to fulfil the particular role of the aircraft.

We have talked glibly about the aircraft being the 'office' for the aircrew and cabin crew throughout this book. It should be clear by now that there are great differences between the office on the ground and the office in the air. The ground-based office offers the following characteristics that make working life comfortable and safe for employees:

- The office is usually large with the opportunity to shift one's horizon from the screen to a wider perspective by looking up. This may include an external view if windows are installed.
- The office provides a well-lit, glare-free environment.
- The office is stable – it does not move.
- Clean air will be provided at appropriate temperature and humidity conditions and a standard will be declared for a number of changes of air per minute.
- The workstation will be designed to meet appropriate health and safety regulations.
- The workstation may be standardised in layout but can be tailored to suit the individual, for example seat height, seat back angle, exact positions of screen, keyboard and mouse.
- The workstation will receive a regular assessment to ensure that it complies with health and safety standards.
- Wherever possible the office environment will be designed to eliminate or minimise such conditions as eye strain, headaches, backache, repetitive strain injury (RSI).

The airborne office differs from this in a number of ways largely determined by the fact that it is installed in a dynamic flying platform:

- The 'office' is usually small, its dimensions limited by the fuselage which is usually of a relatively small diameter and height. The opportunity to shift one's horizon from the screen to a wider perspective by looking up is restricted by these limiting dimensions and by the length of the cabin.

An external view may be possible, but a high-flying aircraft does not often provide an interesting horizon through a small window.

- The office provides a well-lit, glare-free environment, provided that the windows are covered to obstruct sunlight entering.
- The office is not stable – it has six degrees of freedom of motion and changes in some dimensions may induce a feeling of queasiness. In addition to the intentional manoeuvres of the aircraft, turbulence and wind shear introduce rapid rates of change of attitude.
- Clean air will be provided at appropriate temperature and humidity conditions and a standard will be declared for a number of changes of air per minute; however, see Chapter 3 of this book.
- The workstation will be designed to meet appropriate health and safety regulations, but these will be tempered by the need to meet aircraft safety standards with respect to loose articles, security of attachment, as well as the need to meet the aircraft operational requirements.
- The workstation may be standardised in layout but can be tailored to suit the individual, for example seat height, seat back angle, exact positions of screen, keyboard and mouse. This tailoring is restricted on an aircraft because of the need to use standard airline seats and restraints, and by the fact that the keyboard is usually fixed to the desk and a fixed trackball is used rather than a mouse.
- The workstation will receive a regular assessment to ensure that it complies with health and safety standards.
- Wherever possible the office environment will be designed to eliminate or minimise such conditions as eye strain, headaches, backache, repetitive strain injury (RSI). For a bespoke workstation such as this, the assessment will be performed during the appropriate design stages of the aircraft.

9.4 Health Issues

9.4.1 Sight

Eye strain can result from prolonged use of computer screens, even where the lighting is considered to be good. The symptoms of eye strain include dry, itchy or bloodshot eyes and blurred vision. Eye strain can be a common cause of headaches and additionally some people find that looking at a screen whilst moving makes them nauseous.

Ideally the computer monitor should be 18 to 28 inches (45 to 70 cm) from the eyes (about arm's length) and should not face a window or any source of bright light. An anti-glare filter for the screen, or a screen hood, can reduce reflection. Many users strain their eyes because the brightness and contrast levels are badly adjusted and the font size is too small. Computer screens are best used in a room where the lighting level is less than that needed to read a book by – a lamp should be used to provide the directional light required to read papers at a desk.

The most important thing to safeguard the eyes is to take regular screen breaks, where the user can focus on other objects and moisturise the eyes – most people blink less when they are looking at a computer screen. This problem can be worse for contact lens wearers and anyone suffering from an eye complaint such as conjunctivitis will also suffer increased discomfort.

9.4.2 Posture

Wrist and hand pain is one of the most common health problems reported by those regularly using computer keyboards. Poor ergonomics is often the cause – the arm, wrist and hand should be in a straight line when the hands are on the keyboard. A lack of support for the wrists can cause pain and a tingling sensation in the hands and arms. This can result in carpal tunnel syndrome or tendonitis, a painful condition sometimes found in concert pianists. Sitting too far from the keyboard can also lead to strain, as can sitting in a position which is either too low or too high.

9.4.3 Back and Neck Pain

Lower back pain can result from inadequate seating provision – either too high or too low, or on a chair without proper back support. A footrest can put the legs into a more comfortable position for those who are seated for long periods.

Neck pain can be the result of looking at a screen set to one side – it should be placed directly in front of the user and the user should be able to see the screen clearly when holding their head level. Many people adopt a position where the neck is bent 'backwards', however, and the large banks of screens used by some military and commercial operators make it hard to avoid this; the operator has to look 'up' or 'down' or 'sideways' to do their job.

Getting workplace ergonomics right for each individual is not difficult, but aircraft have many different crew members and many different passengers. The big problem with shared equipment in any working environment is that every user has individual needs and shared equipment should be adjustable to suit all shapes and sizes. Cushions, supports and adjustable table and chair heights can help, but are not always available. Where they are available, however, not everyone takes the trouble to make the minor adjustments they need, even though they would probably consider it essential to alter the position of a car seat for comfort and safety.

9.4.4 Vibration

Nausea is the most obvious result of looking at a vibrating computer screen, but prolonged workstation vibration can also cause damage to the wrists and hands.

Despite legislation, some employers and employees are still unaware of the impact on health that a poorly arranged workstation can have. A badly equipped and arranged workstation is a major contributing factor in the development of many work-related upper limb disorders (WRULDs). Conditions can be both short and long term but in most cases cause a lot of avoidable pain, discomfort and stress. Though symptoms such as eye strain, headaches and fatigue-induced stress are usually temporary, other conditions, such as wrist pain, can become debilitating.

The hazards associated with DSE workstations must therefore be properly assessed so that they are adequately equipped and adjustable to suit every user's needs.

9.5 System Implications

It is clearly not feasible to design a commercial aircraft to suit laptop users, since the priority is to ensure that the general public are carried safely and comfortably and permitted to perform domestic things such as reading, eating and listening to entertainment during the flight. Serious users of laptops are still a minority, although that might change as wireless internet access is introduced into aircraft cabins.

Warnings about laptop use should be given by employers and a note in the in-flight magazine would be a good idea.

For military surveillance aircraft, workstations are designed to fit the role and also to fit into the cabin of the type concerned. Most modern workstations will be designed using a 3D modelling design tool such as Catia, and this may well be complemented by the construction of a full-scale mock-up. A full-scale representation of the mission crew compartment may also be constructed in order to carry out crew workload assessments.

System designers should ensure that a robust human factors input included in the design of the military workstation incorporates the following:

- Reach of controls
- Support for arms and wrists
- Seat comfort and adjustment
- Display brightness range
- Appropriate size of fonts and pictorial representations
- Appropriate use of colour
- Provision of ambient and spot lighting.

In the UK, the Health and Safety (Display Screen Equipment) Regulations 1992 require employers to carry out an analysis and assessment of the workstation. This must be carried out by suitably qualified, independent staff.

A risk assessment must be performed on a regular basis. Healthy Working Lives (Healthyworkinglives.com) has produced a DSE Risk Assessment Form that can help with this process. The form describes the assessment process as well as offering advice that could help users to remedy some of the problems that may be encountered in the assessment.

Account must also be taken of daily work routines so that adequate breaks can be incorporated into the working day. This does not necessarily mean a complete break from work, but a break from DSE work. This can be accomplished in surveillance aircraft types by carrying spare crew members or by sharing tasks with other non-DSE tasks. It is better if the task allows for natural breaks, but it is possible to install software that can indicate when it would be appropriate for someone to take a break. Short frequent breaks are better than occasional longer breaks.

Both the employer and employee have a responsibility to ensure correct workstation usage. Appropriate information and training should be provided to users so that they can use the equipment effectively and safely.

Information on eye examinations should also be provided. In most cases an employer is responsible for paying for tests and for basic spectacles if they are required for DSE work.

Airlines should warn all travellers of the health implications of using a laptop in the cabin and companies should alert employees of the risks of laptop use whilst on business.

Reference

Moir, I. and Seabridge, A. (2004) *Design and Development of Aircraft Systems*, John Wiley & Sons, Ltd.

Further Reading

Office Ergonomics Training: www.office-ergo.com (accessed 11 May 2010).

Useful Web Sites

Healthyworkinglives.com

10

Regulation and Control of Risks to Health

10.1 General

The inhabitants of aircraft types tend to fall into two categories: the crew members who fly and operate the aircraft, a class which includes pilots, first officers, cabin crew, navigators, mission crew and flight engineers as appropriate; and people who travel as passengers. Crew members are usually employees of the company operating the aircraft, whilst most passengers have paid for the privilege of travelling. This latter category does not include armed forces personnel who are being transported for military purposes. These various classes of inhabitant are protected by a variety of legislation.

Legislation exists to protect workers or employees, in their workplace. This is often interpreted as protection of office and factory workers, and their working environment is well regulated and governed. The workplace for aircrew and cabin crew is the cabin or cockpit of their aircraft and this is by nature a dynamic environment that is less easy to regulate, but nevertheless impacts on their health. This environment also affects passengers, so any design decisions made in order to comply with legislation will affect them either directly or indirectly.

Many decisions are made during the design and development of a new aircraft which minimise or manage the risk of any circumstances that may be injurious to health. An example route to manage risk in the design and development of an aircraft project is illustrated in Figure 10.1. The actions

Air Travel and Health: A Systems Perspective   Allan Seabridge and Shirley Morgan
© 2010 John Wiley & Sons, Ltd

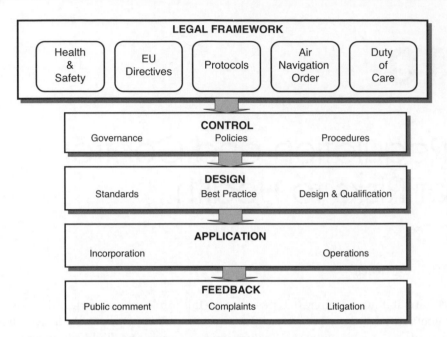

Figure 10.1 Managing health risk in the aircraft project.

described in this route apply to the development of a new project, and they also apply to the definition of changes that arise whilst an aircraft is being operated. The various stages in this route will be described below.

10.2 Legislative Framework

For reasons of currency the legislative framework and the limits that have been used in this book are examples only. It is vital that designers and operators, as well as people seeking to understand the issues, ensure that they meticulously check the latest issues of legislative instruments. This check and understanding must be applied during the lifecycle of any product or project:

- In the concept phase.
- Throughout the development phase.
- Throughout its operational life.
- At its end-of-life disposal.

10.2.1 Health and Safety

The Health and Safety at Work Act in the UK outlines some general conditions that must be met by law to safeguard the health of people in their place of work. All employees have a statutory duty to observe the Act and to demonstrate that they do so. In addition, there are regulations that govern the exposure of workers to specific threats to their health, such as noise, vibration and ionising radiation.

The Act clearly states that an employer has a duty 'to ensure, so far as reasonably practicable' the protection of health and safety and welfare at work. This includes:

- Making the workplace safe and without risk.
- Keeping noise under control.
- Ensuring plant and machinery are safe.
- Providing adequate welfare facilities.
- Communicating health and safety information, instruction and training.

Employees also have duties:

- To take reasonable care for their own health and safety.
- To take reasonable care of the health and safety of others who may be affected by their actions.
- To co-operate with their employer on any health and safety subject.
- Not to interfere or misuse anything provided for health and safety and welfare.

The Health and Safety at Work Act 1974 [3] states:

Section 6

It shall be the duty of any person who undertakes the design or manufacture of any article for use at work to carry out or arrange for the carrying out of any necessary research with a view to the discovery and, so far as is reasonably practicable, the elimination or minimisation of any risks to health or safety to which the design or article may give rise.

Section 10

In any proceedings for an offence under any of the relevant statutory provision consisting of a failure to comply with a duty or requirement to do something so far as is practicable or so far as is reasonably practicable, or to use the best practicable means to do something, it shall be for the accused to prove (as the case may be) that it was not practicable or reasonably practicable to do more than was in fact done to satisfy the duty or requirement, or that there was no better practicable means than was in fact used to satisfy the duty or requirement.

There is a possibility that aircraft operators may be subject to litigation by their aircrew claiming damage to health or impact on their career as a result of using manufactured products in their work. The aircraft operator may then claim that the aircraft manufacturer has made a contribution to this condition as a result of limitations in the design of the product.

The aircraft manufacturer may also be subject to claims under the Consumer Protection Act 1987 if the product is found to be defective and causes death or damage to people or property.

10.2.2 EU Legislation

In the EU some local legislation will be enshrined in European Directives applicable to member states and these directives may become regulations. Typical directives that apply to industry and have an impact on aerospace include:

- Directive 2002/44/EC – Minimum health and safety requirements regarding the exposure of workers to the risks arising from physical agents (vibration). See Chapter 5.
- Directive 96/92 Euratom – Article 42 imposes requirements relating to the assessment and limitation of aircrew members' exposure to cosmic radiation and the provision of information on the effect of cosmic radiation. See Chapter 6.
- Directive 2004/40/EC modified by 2008/46/EC Physical agents (Electromagnetic fields). See Chapter 6.

10.2.2.1 EU Directives

A directive is a legislative act of the EU, which requires member states to achieve a particular result without dictating the means of achieving that result. It can be distinguished from regulations which are self-executing and do not require any implementing measures. Directives normally leave member states with a certain amount of leeway as to the exact rules to be adopted.

Directives can be adopted by means of a variety of legislative procedures depending on their subject matter.

The legal basis for the enactment of directives is Article 249 of the treaty establishing the European Community and, as such, directives only apply within the European Community pillar of the EU:

Article 249

- In order to carry out their task and in accordance with the provisions of this Treaty, the European Parliament acting jointly with the Council, the Council and the Commission shall make regulations and issue directives, take decisions, make recommendations or deliver opinions.
- A regulation shall have general application. It shall be binding in its entirety and directly applicable in all Member States.
- A directive shall be binding, as to the result to be achieved, upon each Member State to which it is addressed, but shall leave to the national authorities the choice of form and methods.
- A decision shall be binding in its entirety upon those to whom it is addressed.
- Recommendations and opinions shall have no binding force.

The Council can delegate legislative authority to the Commission and, depending on the area and the appropriate legislative procedure, both institutions can make laws.

There are Council regulations and Commission regulations. Article 249 does not clearly distinguish between legislative acts and administrative acts, as is normally done in national legal systems.

10.2.2.2 Legal Effect

Directives are only binding on the member states to whom they are addressed, which can be just one member state or a group of them. In practice, however, with the exception of directives related to the Common Agricultural Policy, directives are addressed to all member states.

10.2.2.3 Implementation

When adopted, directives give member states a timetable for the implementation of the intended outcome. Occasionally the laws of a member state may already comply with this outcome and the state involved would only be required to keep its laws in place. But more commonly member states are required to make changes to their laws – commonly referred to as transposition – in order for the directive to be implemented correctly. If a member state fails to pass the required national legislation, or if the national legislation does not adequately comply with the requirements of the directive, the European Commission may initiate legal action against the member state in the European Court of Justice. This may also happen when a member state has transposed a directive in theory but has failed to abide by its provisions in practice.

10.2.2.4 Direct Effect

Notwithstanding the fact that directives were not originally thought to be binding before they were implemented by member states, the European Court of Justice developed the doctrine of direct effect, where unimplemented or badly implemented directives can actually have direct legal force. In *Francovich* v. *Italy* the court found that member states could be liable to pay damages to individuals and companies who had been adversely affected by the non-implementation of a directive.

10.2.2.5 EU Regulations

A regulation is a legislative act of the EU which becomes immediately enforceable as law in all member states simultaneously. Regulations can be distinguished from directives which, at least in principle, need to be transposed into national law. Regulations can be adopted by means of a variety of legislative procedures depending on their subject matter.

In some sense, regulations are equivalent to 'Acts of Parliament', in that what they say is law, and they do not need to be mediated into national law by means of implementing measures. As such, regulations constitute one of the most powerful forms of EU law and a great deal of care is required in their drafting and formulation.

When a regulation comes into force it overrides all national laws dealing with the same subject matter and subsequent national legislation must be consistent with and made in the light of the regulation. While member states are prohibited from obscuring the direct effect of regulations, it is common practice to pass legislation dealing with consequential matters arising from the coming into force of a regulation.

10.2.3 Environmental Legislation

Environmental concerns have led to restrictions on the use of certain materials to protect the environment and to reduce the risk to humans. The Montreal

Protocol has banned the use of chloro-fluoro-carbon (CFC) compounds to protect the ozone layer. This has led to the restriction on the use of CFCs and halons in refrigerants and in fire-suppressant fluids. Other materials have been banned or restricted in their use because of human health concerns. Although this is mainly to protect people involved in the manufacture of the aircraft, a source of passenger and crew contamination has nevertheless been removed. Examples of banned or restricted hazardous materials include asbestos, lead from solder, cadmium plating, beryllium alloys, VOC-based paints and cleaning fluids such as tri-chloro-ethylene.

Because of the long lifecycle of aircraft products, some of these materials will inevitably be present in some older aircraft but will not be incorporated in new designs. They may present a hazard if an aircraft crashes and causes contamination of the crash site and surrounding area. In the disposal phase of the aircraft care must be taken to ensure that materials are correctly handled and disposed of.

10.2.4 Air Navigation Order

Some legislation is incorporated into the ANO with the specific intention of protecting aircrew and cabin crew. See Chapter 6 for an example.

10.2.5 Duty of Care

The manufacturer of a product has a duty of care to its employees, so that those people employed to design and build the aircraft are protected. There is also a duty of care extended to the operators and users of the aircraft to ensure that no users are injured or affected by the design of the product in service.

The aircraft operator also has a duty of care to its employees – those people maintaining and servicing the aircraft as well as those crew members who operate the aircraft.

10.3 Summary of Legal Threats

In the event that products supplied by an aircraft manufacturer are found to be unsafe or harmful to those operating them, then there are several potential sanctions that may apply.

10.3.1 Criminal Prosecution

This is likely if it can be shown that the manufacturer is in breach of relevant legislation, particularly the Health and Safety at Work Act 1974, but also a whole series of safety-related legislation, most of which is from EU directives. The penalties that the company would suffer will generally be fines – ranging from relatively small amounts to significant sums of money. There is also the possibility of corporate manslaughter charges arising out of negligence leading to the death of individuals. The first corporate manslaughter trial in the UK began in February 2010.

10.3.2 Civil Lawsuits

If individuals suffer injury as a result of the use of a product and that injury can be shown to have been caused by a defect within the product, then the manufacturer faces the prospect of being sued by that individual for damages. The amount payable will depend upon the severity of the injury but, in any event, if the manufacturer is held liable it will be required to pay the legal costs of all parties involved, correct the defect in the product, possibly recalling existing items for repair. Customers may also take legal action against manufacturers if the products that are supplied or maintained are defective. Again, the penalty for this will be damages and significant legal costs.

10.3.3 Customer/Public Relations

If an aircraft manufacturer develops a reputation for supplying products which are inherently unsafe and which lead to users suffering harm, then customers are less likely to purchase from that manufacturer, with the consequent impact on profits and shareholder value. People have been known to shun particular airlines because of their safety record.

Manufacturers need to ensure that the products they supply are as safe as possible given all the circumstances and that they continue to evaluate and minimise risk wherever possible. Failure to do this can have significant and far-reaching consequences.

10.4 Issues Arising

The responsible manufacturer of aircraft will need to address certain issues including the following:

1. Exactly what duty of care does the aircraft manufacturer owe to its customers and employees in respect of the products (and services) that it manufactures and supplies?

2. How does the aircraft manufacturer determine the levels of risk contained within the products that it manufactures and supplies?
3. How far does the aircraft manufacturer have to go to develop processes to prevent problems from arising?
4. Once a product is in the marketplace, what is the aircraft manufacturer's responsibility for it? What are the ongoing obligations once the product is in use?
5. What level of competence does the aircraft manufacturer need to maintain within the company to carry out research and development in the areas of product safety? This is especially relevant to the initial portion of Section 6 of the Health and Safety Act which states that:

> It shall be the duty of any person who undertakes the design or manufacture of any article for use at work to carry out or arrange for the carrying out of any necessary research with a view to the discovery ... of any risks to health or safety to which the design or article may give rise.

6. Allied to 5 above, what responsibility does the aircraft manufacturer have to monitor changes in legislation and update its products accordingly?

11

The Design Process

The control of risk described in this chapter follows the route shown in Figure 10.1 in the previous chapter, using inputs as required from the legal framework.

11.1 Control of Risk

In the design and development of an aerospace product, great care will be taken to ensure that the product is fit for purpose and complies with all legislation existing at the start of design. The product may be modified throughout its life to maintain compliance, although the application of changes to achieve this may well lag behind the current state of legislation. Most industries will make use of a process that is repeatable and can be used as justification of good practice in operation.

11.1.1 Governance

Good and robust design is expected of any manufacturer of aircraft products and this is usually enforced by sound governance in the organisation. Governance includes the selection and training of qualified engineers, the deployment of sound process and procedures, the maintenance of an organisational structure that incorporates independent verification of design and application of standards, and an effective mechanism for collecting design and qualification data and verifying their correctness and applicability to the product.

Air Travel and Health: A Systems Perspective   Allan Seabridge and Shirley Morgan
© 2010 John Wiley & Sons, Ltd

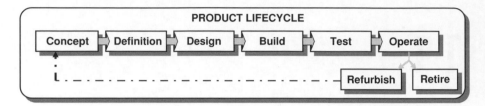

Figure 11.1 A simplified design and development lifecycle.

This is most effective when practised by an organisation independent of the project teams, and further refined by an independent airworthiness authority.

11.1.2 Company Policy

Companies will normally have policies for specific issues to ensure that there is a common understanding in the various departments on how to address these, and also to give the customer confidence in the organisation's ability to design a robust product.

11.1.3 Company Procedures

Where necessary a company will enact its policies in the form of procedures to ensure that the organisation maintains a consistent approach to designing, building, testing and qualifying its products.

11.2 Design

11.2.1 Standards

Standards will often be requested by the customer of the aircraft, particularly in the military field, where a choice is available between the US Military Standard (MIL-STD), the UK Defence Standard (Def-Stan) or British Standard (BS) and French standards (Air Règlements), for example.

Local standards will be consulted for materials and components by suppliers. The standards will be applied at the most recent issue in existence at the beginning of the project and must be maintained with any changes in issue referred to the project teams.

11.2.2 Good Practice

Good practice includes the appropriate use of previous design experience on similar projects mixed with knowledge of new technology and processes. Good practice may be stored as intellectual property and is often vested in skilled and experienced staff, backed up with design manuals and experience databases.

11.2.3 Design and Qualification Process

The task facing designers of aircraft is to understand the requirements of the numerous and disparate aircraft systems that need to be designed and developed into an integrated system solution to ensure that the aircraft is equipped to perform its stated tasks. To develop such a system from a customer's requirement through to implementation requires a discipline that will enable people to apply their skills and experience in a rigorous and consistent manner.

It is important to recognise that the product moves through a number of stages in a lifecycle that covers initial concept, design and development and in-service operation by a customer until the product is no longer required. In the case of an aircraft this entire lifecycle is generally about 25 years, and with mid-life updates and refurbishment may exceed 50 years, as is the case with some aircraft in service today. This lifecycle is shown in a very simplified form in Figure 11.1 (Moir and Seabridge, 2004). Because the lifecycle for modern aircraft can exceed 50 years, there are many aircraft flying today whose design concepts can be traced back for half a century or even longer.

Inevitably in such a prolonged lifecycle there will be issues of currency of technology, obsolescence, changing requirements, application of different skills and processes and changing legislation. Even the initial development phases before product design is sufficiently mature to commit to production are now longer than some technology life spans – in other words, 'new' technology may be obsolete before it can even be used, yet alone stay in service for 25 years. As well as technology advancing in the aircraft lifecycle, so does legislation, and it may well be that an aircraft that entered service perfectly suited to its task becomes out of step with legislation.

Companies that design aircraft adopt a disciplined approach to design and development in order to produce an aircraft that is fit for purpose and that meets the requirements of both the purchasers and the legislative framework. This process is described briefly to provide readers with an understanding of what is involved in designing, building, testing and

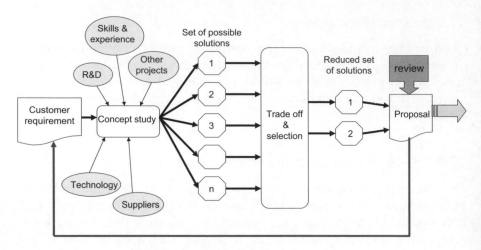

Figure 11.2 The concept phase process.

operating an aircraft throughout its life to meet the customer requirements and existing legislation.

To give this context some firm links to the issues discussed in the book, a brief description of engineering considerations relevant to the health of aircraft occupants is provided at the end of each lifecycle stage description.

11.2.3.1 Concept Phase

Figure 11.2 illustrates the key engineering activities associated with this phase of the lifecycle.

The concept phase is about understanding the customer's emerging needs and arriving at a conceptual model of a solution to address those needs. The customer continuously assesses its current assets and determines their effectiveness to meet future requirements. The need for a new military system can arise from a change in the local or world political scene, or a perceived changing threat that requires an adaptation to defence policy. The need for a new commercial aircraft may be driven by changes in national and global travel patterns resulting from business or leisure traveller demands.

The customer requirements will be made available to industry so that solutions can be developed specifically for that purpose, or that can be adapted from the current R&D base. This is an ideal opportunity for industry to discuss and understand the requirements to the mutual benefit of the customer and its industrial suppliers, to understand the implications of providing a fully compliant solution or one which is aggressive and sympathetic to marketplace requirements. Not all R&D is driven by the customer, nor is it all customer funded. Industry will, as part of its forward-looking strategy, seek to identify and carry out speculative, self-funded research. This may be to support current projects or to reduce the risk of proposing innovative solutions, but may also be non-project-related 'blue skies' research.

Typical considerations at this phase are:

- Establishing and understanding the primary role and functions of the required system.
- Establishing and understanding desired performance and market drivers such as:
  - Range
  - Endurance
  - Routes or missions
  - Technology baseline
  - Operational roles
  - Number of passengers
  - Mass, number and type of weapons (for military types)
  - Availability and despatch reliability
  - Fleet size to perform the role or satisfy the routes
  - Purchase budget available
  - Operating or through-life costs
  - Market size and export potential
  - Customer preference.
- Establishing confidence that the requirement can be met within acceptable commercial or technological risk.
- Developing an understanding of a solution that can be manufactured. This will lead to proposed aircraft shapes, interior and exterior configurations, together with preliminary system architectures.

This phase relies considerably on the skills and experience of the engineering design staff. They will look back at previous solutions and the lessons learned from in-service operations, examine new technology and suitable applications and relate this to any new requirements.

The output from this phase is usually in the form of reports, drawings, mathematical models or brochures. The customer may use these to refine its initial requirement by incorporating new information or by taking into account the risks identified. As implied by the title of the phase, the output is a conceptual design and does not necessarily guarantee that the proposed system is optimal or that it could be manufactured. The output is intended to be sufficient for the customer and industry to agree jointly to move on to a more detailed definition phase. In fact the outcome may be a number of potential solutions from which a choice has to be made.

This phase is focused on establishing confidence that the requirement can be met within the bounds of acceptable commercial or technological risk. This process allows possible vendors to establish their technical and other capabilities and represents an opportunity for the aircraft company to assess and quantify the relative strengths of competitors and also to capture mature technology.

Health Considerations

The concept phase addresses such issues as the flight profile of the aircraft and its operating environment. This will determine under what conditions the aircraft is expected to operate in terms of performance, atmospheric and climatic. At this phase it is important for systems integrators to understand the legislation and the technology available to provide inputs into top-level specifications and to convince the customer that the emerging concept will be fit for purpose.

For the ECS, for example, it determines the volume of the inhabited space of the aircraft, the number of crew members and passengers that need to be kept in comfortable conditions, as well as the systems on the aircraft that need to cooled. All these enable ECS designers to estimate the amount of heating or cooling that is required to maintain the correct cabin conditions in the areas of the world in which the aircraft will operate. This determines the volume and pressure of air required from the engine (or alternative source) and the size of the cooling machine. R&D activity will examine the current scope of airborne pollutants and biological or viral attacks in order to determine what filtration will be required.

11.2.3.2 Definition Phase

Figure 11.3 illustrates the key engineering activities associated with this phase of the lifecycle.

The customer will usually consolidate all the information gathered during the concept phase to firm up its requirements. This often results in the issue of a specification or a Request For Proposal (RFP). This allows industry to develop its concepts into a firm definition, to evaluate the technical,

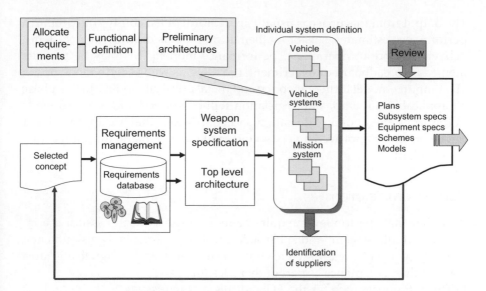

Figure 11.3   The definition phase process.

technological and commercial risks, and to examine the feasibility of completing the design and moving to a series production solution. Typical considerations at this stage are:

- Developing the concept into a firm definition of a solution.
- Developing system architectures and system configurations.
- Re-evaluating the supplier base to establish what equipment, components and materials are available, or may be needed to support the emerging design.
- Defining physical and installation characteristics and interface requirements.
- Developing models of the individual systems.
- Quantifying key systems performance measures such as:
    - Mass
    - Volume
    - Growth capability
    - Range/endurance.
- Identifying risk and introducing mitigation plans.
- Selecting and confirming appropriate technology.

The output from this phase is usually in the form of feasibility study reports, performance estimates, sets of mathematical models of individual systems' behaviours and an operational performance model. This may be complemented by breadboard or experimental models, as well as with mock-ups in 3D computer model form or wooden and metal physical models. In this phase all applicable standards and legislation will be applied and a record made of the applicability or any deviation and concessions that are approved. For deviations to be approved a very sound case must be made for not meeting the standards.

Health Considerations

It is vital that the top-level requirements to address health issues are incorporated into system- and subsystem-level requirements. Legislation and technology must be reassessed to confirm suitability and design teams must confirm their acceptance as part of their design.

The definition phase for the ECS addresses such issues as the volume of the cabin and the number of changes of air required per minute to achieve comfortable conditions, the size and routing of air ducts and vents, the type and number of air filters required. The density and range of airborne pollutants will be determined in order to specify filters and the applicability of standards and legislation will be checked to confirm that the specification is correct.

Consideration will also be given to all other factors that may affect health. The radiation risk will be assessed based on the performance of the aircraft in terms of duration and height; the cabin environment will be assessed and the crew seating conditions will also be examined. Decisions will be made about whether solutions are available in the technical definition or whether procedural limits need to be incorporated on crew and passenger flight-hours.

11.2.3.3 Design Phase

Figure 11.4 illustrates the key engineering activities associated with this phase of the lifecycle.

If the outcome of the definition phase is successful and a decision is made to proceed further, then industry embarks on the design phase. Design takes the definition phase architectures and schemes and refines them to a standard that can be manufactured.

Detailed design of the airframe ensures that the structure is aerodynamically sound, of appropriate strength and able to carry the crew, passengers,

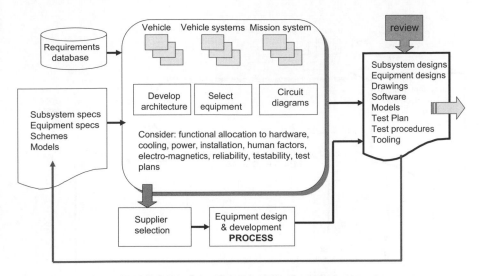

Figure 11.4   The design phase process.

fuel and systems that are required to turn it into a useful product. As part of the detailed design, attention must be paid to the mandated rules and regulations which apply to the design of an aircraft or to airborne equipment. Three-dimensional solid modelling tools are used to produce the design drawings in a format that can be used to drive machine tools to manufacture parts for assembly.

Systems are developed beyond the block diagram architectural drawings into detailed wiring diagrams. Suppliers of bought-in equipment and components are selected and they become an inherent part of the process, in starting to design equipment that can be used in the aircraft and systems.

Health Considerations

In this phase it is important to ensure that all relevant supplier specifications have included the appropriate requirements and that the expertise of suppliers is being used to best effect.

The design phase for the ECS, for example, turns the preliminary design ideas into hard design drawings. Specifications for equipment will be compiled for items such as the air cycle cooling pack, the air ducts, air vents and filters, and these will be passed to suppliers who will apply their own specialist knowledge to design equipment or to offer off-the-shelf items. The

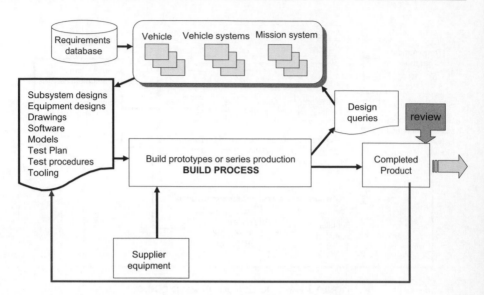

Figure 11.5 The build phase process.

suppliers usually have a wide range of customers and will have gained experience both from other aviation projects and from projects outside the aerospace sector. This also applies to all other systems that may have an impact on health. If a procedural solution is required then an appropriate statement will be incorporated into the operating manuals and flight clearance documentation.

11.2.3.4 Build Phase

Figure 11.5 illustrates the key engineering activities associated with this phase of the lifecycle.

The aircraft is manufactured to the drawings and data issued by design. This includes the fabrication of detailed subassemblies and their progressive build-up, or final assembly, into a complete airframe together with the installation of pipes, ducts, wiring harnesses and equipment. The main systems engineering support to this phase is to provide a service to manufacturing in answering queries in instances where the solution cannot be achieved in practice or in an economical manner from a quantity production viewpoint. Prompt and effective answers at the early stages of build can reduce the probability of errors appearing in quantity production.

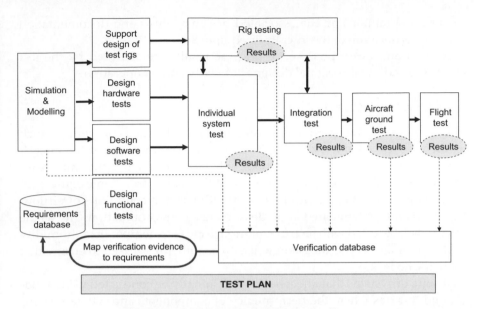

Figure 11.6   The test phase process.

Health Considerations

The build phase for the ECS will examine the technology choices available, the current range of fully developed products, and will decide how to proceed – either to develop a new solution or to use an existing product. There will be a temptation to reuse existing parts and well-proven systems already in use, but new and emerging contamination threats will be recognised and used to develop a suitable filtration mechanism which can be offered to the aircraft manufacturer for installation into the ECS.

11.2.3.5 Test Phase

Figure 11.6 illustrates the key engineering activities associated with this phase of the lifecycle.

The aircraft and its components are subject to a rigorous test programme to verify their fitness for purpose. This programme includes testing and progressive integration of equipment, components, subassemblies on various test facilities and rigs, and eventually into the complete aircraft. Functional testing of systems on the ground and during flight trials verifies that the performance and operation of the equipment is as specified. Conclusion of the test programme and the associated design analysis and documentation leads to certification of the aircraft or equipment.

At each and every phase of testing the results are analysed by the test engineers and by the original designer to verify correct operation.

Health Considerations

The test phase for the ECS includes testing of its individual components such as the air cycle cooling pack and its incorporation into a rig at the aircraft manufacturer's site. Here the ECS pack will be connected to representative air ducts and simulated heat load to represent the cabin and avionic equipment heat dissipation. This rig will be tested with varying conditions of airflow, heat load and contaminated air to demonstrate correct operation. The whole aircraft will be tested with instrumentation to measure temperature, humidity and air cleanliness and the cabin will also be fully pressurised to ensure that there are no leaks.

Testing of systems that affect health will initially be conducted using models and test rigs where the performance of components and whole systems will be measured and the results compared with standards. It is rare to expose human beings to testing, partly because of the risk to health but mainly because of the variability of humans and the difficulty in repeating experiments with precision.

11.2.3.6 Operate Phase

Figure 11.7 illustrates the key engineering activities associated with this phase of the lifecycle.

During this phase the customer is operating the aircraft on a daily basis. Its performance will be monitored by means of a formal defect reporting process, so that any defects or faults that arise are analysed by the manufacturer. It is possible to attribute causes to faults such as random component failures, operator mishandling or design errors. The aircraft manufacturer and its suppliers are expected to participate in the attribution and rectification of problems arising during aircraft operations, as determined by the contract. The operator will also perform routine maintenance to maintain the aircraft safe to fly and will also contract out cleaning operations between flights.

The collection of information during the operational life of the aircraft may result in changes to correct defects, introduce improvements or introduce corrections resulting from customer feedback. The request for a change is formalised and will go through a process similar to the design process to ensure that any changes are introduced correctly.

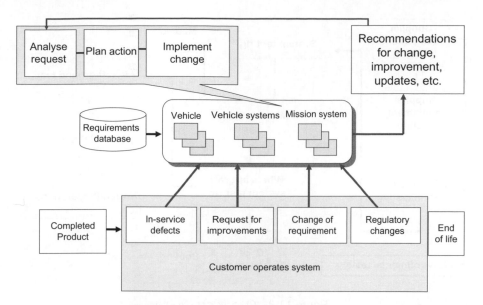

Figure 11.7 The operate phase process.

Health Considerations

The operate phase for the ECS addresses such issues as cleanliness of the air ducts and vents to monitor the build-up of dust or moulds in the system. The filters in the system will also be changed and cleaned or replaced on a regular basis.

A hypothetical change may arise from the discovery of a virus that is able to pass through the currently installed filter. Confirmation of the virus and its size, and the fact that it is an established new source of infection, will result in a request for a change to the filter mesh diameter. A design will be produced and approved, tests will be conducted to prove its efficacy, and new filter assemblies will be manufactured for installation into the operational fleet and any aircraft on the production line.

11.2.3.7 Qualification

The results of all the preceding phases are collated to form a data set that enables the design and the test results to be scrutinised to establish that the product is fit for purpose. This documentation forms a record of the design and is also available to support any incidents reported during operation of

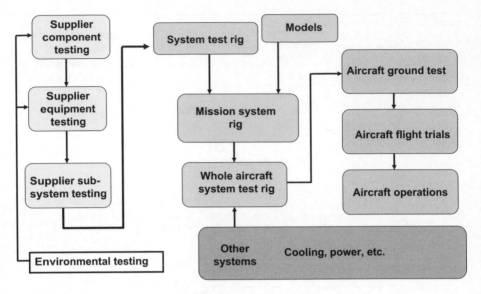

Figure 11.8 Qualification route.

the aircraft, as illustrated in Figure 11.8. It is therefore possible to trace the design and the decisions made during the entire development process.

A more realistic lifecycle can be seen in Figure 11.9, in which the operate phase is very much longer than the development phases. It is not unusual for some aircraft types to remain operational for up to 50 years, often with a number of updates to their systems. During this time technology maintains its steady march, people become aware of risks to health and the ready availability of instant communications means that health issues are widely divulged to the public. Legislation also advances and will affect new projects designed to enter service at the end of an existing project's life, but less likely to be retrofitted into existing aircraft. This is often perceived as a failure on the part of industry and legislators to react to public demands.

11.3 Application

Adherence to this design process results in an aircraft that is fit for purpose – it will do the job for which it was designed without causing damage to its inhabitants.

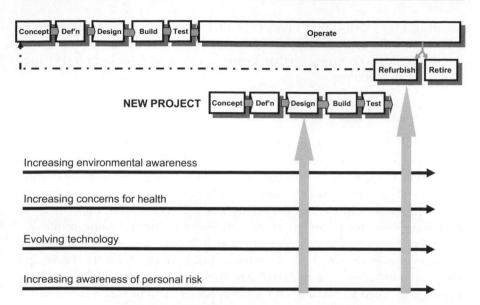

Figure 11.9 A more realistic lifecycle duration.

11.3.1 Incorporation

The designed and certified product is incorporated in to the operator's fleet. It is usual that the core of the aircraft – the airframe and the major systems – will be common, but different operators may demand different engines and will certainly want different interior furnishings and decoration, as well as the exterior paint finish and logos.

11.3.2 Operations

The aircraft will then enter service performing normal operations on declared routes. It will also be subject to routine maintenance operations and any faults that occur will be rectified, with commonly occurring faults referred back to the aircraft manufacture for design investigation. Repairs will also be performed should any damage occur. None of these activities should conflict with the basic design standard.

11.4 Feedback

People do complain about their travel experiences. The Sunday newspaper travel supplements contain letters complaining about the service of hotels and travel agents as well as the providers of air, land and sea transportation. If the issues are serious or much reported then research will usually be conducted by the provider concerned, so that it can provide a basis for technical change to resolve the problem.

11.4.1 Public Comment

The travelling public make good use of a number of publicly accessible sources to air their grievances. The Letters to the Editor page was once the main route to express concerns but now the travel supplements publish complaints and in some cases take action on behalf of the readers. This makes travel operators and airlines the immediate subject of public judgement. In addition, web sites of organisations representing the travelling public are open for comment and people also publish 'blogs'. These and a variety of other sources of opinion have been used during research for this book, and the vast array of 'source material' has the effect of making adverse travel experiences open to anyone who actively seeks them out. The end result is that the internet and media are full of conflicting views.

11.4.2 Complaints

Formal complaints to operators are made and the public expect a result in their favour. If this does not happen then they may write to newspaper travel correspondents or organisations that represent travellers, who will often take up good cases.

In the case of crew members, the route for a formal complaint is direct to their employer or to their trade union representative.

11.4.3 Litigation

Legal action, once initiated, can take some time to arrive at a satisfactory conclusion unless the issue is beyond doubt. Examination of the evidence sometimes uncovers many conflicting views and is often inconclusive. From the passenger's perspective, consistency in rulings involving air travel can seem hard to find.

There have been many instances of passengers – sometimes acting as a group – attempting to sue airlines for damages. Most cases involve passengers trying to get compensation after suffering from DVT, or from other health problems attributed to cramped conditions on the aircraft, but there have been instances where passengers injured by items falling from the overhead lockers have filed a claim for negligence. Whatever the circumstances, the outcome in all cases is far from certain.

In 2005 The House of Lords ruled against eight people affected by DVT, who had been fighting for four years to be allowed to sue several airlines, claiming that DVT is caused by cramped seats. The airlines had denied liability, saying this would not meet the legal definition of an accident.

In the same year the High Court of Australia ruled that an international airline passenger could not pursue a claim against an airline for DVT allegedly contracted during a flight because the passenger could not prove the DVT was caused by an 'accident' within the terms of the Warsaw Convention (*Povey* v. *Qantas Airways and British Airways*).

In 2002, however, a woman passenger won £13 000 ($19 500) compensation after she was 'squashed' by an obese person who sat next to her on a transatlantic flight.

Amongst the many 'reasonable' claims for reparation are some incredible ones which seem to have little chance of success but still add fuel to the debate on air travel, health and liability. For example, in 2004 a couple from the US complained that cramped conditions on their flight to Europe caused them both to suffer from bad backs on arrival. In the man's case, his subsequent back and leg pain resulted in his tripping at an Italian cathedral, breaking his nose and smashing his teeth. It had been several days since the flight, but the couple blamed the airline and were prepared to sue to win damages (http://chicagoist.com/2004/05/17/couple_sues_over_more_legroom.php).

Incredibly, on 17 December 2008 the *Telegraph* reported the case of a Japanese man who sued his carrier for making him so drunk that he slapped his wife as they left the terminal. He was arrested for assault but said the airline was irresponsible for making him drunk by serving brandies at 20-minute intervals. The airline said any suggestion that it could be held responsible for a passenger's assault on his wife was 'without merit'.

Reference

Moir, I. and Seabridge, A. (2004) *Design and Development of Aircraft Systems*, John Wiley & Sons, Ltd.

12

Summary and Conclusions

There are factors that make an aircraft different from all other forms of transportation. The aircraft operates in an atmosphere in which the air density is lower than at sea level, there is less oxygen and it is very cold. The aircraft is therefore critically dependent on its engines for providing a source of air that can be used to provide a controlled temperature and pressure environment in the cabin and also allow passengers to breathe. The passengers are totally enclosed, they can only leave the environment when the aircraft has landed and is parked on the ramp, they have limited freedom of movement and they have to sit, eat and sleep in close proximity to strangers – an unusual circumstance for most people. They have also been exposed to media accounts of illnesses, syndromes and viruses and may be in a state of anxiety. They are also exposed to real contaminants in the cabin that can lead to short-term or chronic ill health.

In this book a number of potential contributory factors to ill health have been examined. We have presented lots of evidence, some anecdotes and some well-established myths. We must leave it to the readers to make up their own minds in many cases, given the conflicting messages that research sometimes throws up. If some views are somewhat sceptical it is because the sum of experience and the number of sufferers is small in statistical terms. It can be almost impossible to compare 'like for like' situations when people report a wide range of imprecise symptoms. Nevertheless, there are some

Air Travel and Health: A Systems Perspective   Allan Seabridge and Shirley Morgan
© 2010 John Wiley & Sons, Ltd

lessons to be learned by designers and operators and it is hoped that some of these lessons can be applied to reduce the risk to individuals.

Each of the factors considered in relation to flying raised its own set of symptoms and it was clear that there was some commonality. Table 12.1 later in this chapter lists all the symptoms quoted by sufferers against the various factors described in each chapter. It is also clear that there are people who claim to suffer from factors in the normal ground environment, and their symptoms are also similar. They are included in the table to highlight similarities.

There is a dilemma here in that the population of people flying on a daily basis is high, but the number of people who suffer anything more than mild irritation is very low. This is not to say that the sufferer of a serious condition is imagining their ill health, rather that the cause may be other than aerospace related, or may be a combination of conditions in their home or work environment, exacerbated by flying.

It is probably this imbalance that leads to a lack of investment in trying to find solutions to conditions whose cause is unknown. Research will be performed into symptoms and possible causes, but it takes time to determine a statistically significant result upon which industry is able to start serious investigations into technical solutions. Scientists and engineers need a firm requirement on which to base their work, a situation often seen by the media and the public as prevarication, and this is reflected in comments on various web sites by people who feel that they are not being taken seriously.

12.1 Integration Effects

Remember that the inhabitants of an aircraft are not subjected to the conditions described singly, but rather in combination. The topics are described as single items of study in most research, and many sufferers tend to attribute their symptoms to a single cause. However, it is clear that in systems engineering terminology the aircraft is acting as an integrating medium, causing all of its occupants to be subjected in equal measure to the internal and external environment in any one flight, as illustrated in Figure 1.2 of Chapter 1. What causes a difference between passengers and crews is that passengers are usually exposed to this integrated set of conditions on an occasional and short-term basis, whereas the crew are exposed frequently and for long durations as part of their job.

Table 12.1 Symptoms exhibited by airline passengers compared with people exposed to external conditions – two matches or more.

Symptom	Jet lag	Fear of flying	Air Quality	WiFi[1]	EM[2]	New car[3]	Cleaning Fluid[4]	Sick building[5]	ME[6]	No of matches
Light Headed/dizzy	X	X	X	X	X	X	X	X	X	9
Headache	X		X	X	X	X	X	X	X	8
Rash/Burning face			X	X	X	X	X	X	X	7
Nausea			X	X	X	X	X	X	X	7
Tired/fatigue	X		X	X	X	X		X	X	7
Breathless/tight chest			X	X	X	X		X	X	6
Poor memory/word recall			X	X	X	X		X	X	6
Lack of concentration		X	X	X	X			X	X	6
Eye irritation			X		X	X	X	X	X	6
Depression			X	X	X		X		X	5
Palpitations		X	X	X	X				X	5
Digestive problems	X		X	X	X				X	5
Diarrhoea/stomach upset			X				X	X	X	5
Panic/hysteria/confusion		X	X		X			X	X	5
Insomnia	X		X	X	X				X	5
Flu symptoms			X	X		X			X	4
Tingling sensation			X	X	X				X	4
Balance impaired	X		X		X				X	4
Dry/sore throat			X				X	X	X	4
Vision impaired			X		X				X	3
Drowsiness			X			X	X			3
Cough/wheezing			X			X		X		3
Irritable				X	X			X		3
Emesis/vomiting		X				X	X			3
Muscle spasm/shake			X		X				X	3
Detached mood			X	X					X	3
Tinnitus/hearing loss				X	X				X	3
Night sweats			X						X	2
Loss of feeling/numbness				X					X	2
Hair loss				X					X	2
Fever/sweating								X	X	2
Chest pain								X	X	2
Muscular pain								X	X	2
Cough								X	X	2

(*Continued*)

Table 12.1   (Continued)

Symptom	Jet lag	Fear of lying	Air Quality	WiFi[1]	EM[2]	New car[3]	Cleaning Fluid[4]	Sick building[5]	ME[6]	No of matches
Loss of coordination						X			X	2
Nosebleed/epistaxis				X		X				2
High blood pressure				X	X					2
Irregular heartbeat					X			X		2
Mood changes								X	X	2
Confusion								X	X	2
Sneezing							X	X		2

Sources:
[1] Adams (2008), Beschizza (2006), Philips (2009), *South Wales Evening Post* (2009), Moore (2007), *The Times* (2009).
[2] Graham (2005), US Environmental Protection Agency (1994), Lab News (2009), Air Quality Sciences (2006), Sixwise.com (2005), Article Alley (2005).
[3] Diagnoseme.com (2009), The Environmental Illness Resource (2009), Natural Matters (2006), Chemical Free (2009), EMF Blues (2009).
[4] BUPA (2004), Sixwise.com (2005), *The Ecologist* (2009), *Guardian* (2004), Institute of Environmental Health (1999), Babyworld (www.babyworld.co.uk).
[5] WHO (1983), Environmental Illness Resource (2009), Safe Workers (2009), Medicinenet.com.

Another, less visible, impact of integrated systems is that changes in one system can impact on another, apparently unrelated system. An example of this is the uprating of engines to improve performance, which can lead to higher noise levels in the cockpit or cabin. Another example is the application of cost-saving measures, which can lead to crews reducing the air-conditioning when the aircraft is parked in order to save on fuel costs. This may lead to passenger discomfort and an increased risk of biological cross-contamination.

12.2 Predisposition

There are a number of factors to be taken into account when designing an aircraft to be fit for purpose and when operating an aircraft to carry customers or to fulfil a military requirement. But the impact on the health of occupants is affected considerably by circumstances beyond the control of the designer or operator. This needs to be considered in any attempts to initiate litigation or in preparing a defence against such an action. Predisposition is a factor in everything covered in this book and, although there are undoubtedly trends, illness and/or pain thresholds are very subjective issues, making it hard to compare one person's experience with another. All these factors must be balanced in judging any claims of ill health related solely to flying.

12.2.1 The Travel Experience

The whole travel experience may contribute more to any ill effects than the flight itself. This has already been mentioned in Chapter 1, where a list of events encountered in any return journey can leave an impression that even the most hardened traveller dreads repeating.

People may be inclined to forget that they have lifted and carried bags to the car or cab, carried them through the airport, queued at check-in and security, maybe carried children and carry-on bags, and then sunk into what they hope is a comfortable seat. They will then try to sit still for hours, eat, doze, sleep and then repeat the bag carrying on arrival.

12.2.2 Genetic Factors

Some travellers may be genetically predisposed to some ailments; they may know this from their family history, or they may not yet be aware. Typical conditions which have been encountered in this book include some forms of cancer, respiratory illnesses, heart and circulatory diseases and anxiety.

12.2.3 Public Health

Before boarding the aircraft at their home airport, travellers are exposed to an environment that is beyond their control and in which there are diseases and viruses in circulation. To a certain extent they may have some natural resistance to some of these factors, because that is where they live. On arrival at a foreign airport the environment is different and may contain sources of disease that the traveller has not encountered previously. It is easy to blame a bad stomach on the flight, rather than the sandwich eaten at the airport.

12.3 Domestic Circumstances

What people do in their domestic lives may also predispose them to certain conditions. Some people will carry on with their domestic and work lives right until their trip starts, and they will carry on with their holiday activities until their return trip starts. Thus there is often little break in people's active lives other than the travel experience. This can have a complicating impact on some health conditions that make it difficult to determine what effect the flight experience had.

12.3.1 Carrying and Lifting

Straining the back or back muscles doing heavy lifting at home can cloud the issue of back-pain-related disability claims. Heavy lifting can include shopping, carrying baggage or even picking up children.

12.3.2 Do-It-Yourself (DIY)

DIY can involve activities that include bending, stretching and lifting, and may occur over several days, certainly over many hours. Gardening can involve strenuous digging or bending and this can lead to back and muscle pain, which may not appear immediately. Lifting sacks, bricks and logs or breaking stones has a similar effect. Decorating also uses muscles that may otherwise never be exercised.

Doing this immediately before a trip may complicate the issues of carrying bags, standing in queues and sitting for hours in an unfamiliar and uncomfortable seat.

12.3.3 Noisy Pursuits

Motorsports, shooting, even listening to music, can make a contribution to the daily noise dose. Common causes of hearing-related disorders include loud music from clubs, MP3 players with in-earphones and loud car audio systems.

12.3.4 Lifestyle Factors

The reputation of aircrew as a sunbathing partying set is undoubtedly a complicating factor. Whether true or not, there have been examples of experts ascribing skin cancer in cabin crew members as a direct result of their lifestyle. An example is the air hostess who died of skin cancer because her job gave her the opportunity to sunbathe. Flying did not kill her, but it was seen as a contributory factor. Leisure activities that include diving can also affect the gases in the blood to the extent that the 'bends' may become a significant risk.

12.3.5 Water Sports

Activities that involve diving to any reasonable depth can affect the concentration of gases dissolved in the blood and can lead to the occurrence of the bends in the presence of rapid changes of pressurisation in the cabin.

12.3.6 Obesity

Obesity is becoming a fact of life in some cultures and it can contribute to the general discomfort of travel by making it difficult to negotiate the cabin aisles and to get comfortable in the seat. People with severe obstructive sleep apnoea (OSA) on commercial flights may have a greater risk of cardiac stress than healthy people, according to research. Their risk of serious heart problems is increased because their breathing is shallow and irregular during sleep, which means their body has to work extra hard in the pressurised cabin environment. OSA is becoming more common because it is often associated with obesity, where extra body fat can put a strain on the muscles in the throat (Science Daily, 2008; NHS Choices, nhs.uk/conditions; thaindian.com/newsportal).

12.4 Comparison with General Public Health Conditions

There are a number of conditions that some people claim to suffer from in the natural and human-made environment that seem to present similar symptoms to those encountered in flight. Typical conditions include the following.

12.4.1 Sick Building Syndrome

Sick building syndrome has been called a 'definite entity' by the Health and Safety Executive (HSE) and the WHO. Sufferers report a variety of symptoms that affect them while in a particular building. Symptoms such as a stuffy or runny nose are characteristic of an allergic reaction whilst others, such as extreme lethargy, suggest a different cause.

Research carried out by one US company specialising in remedying 'sick' buildings showed that allergenic fungi were the main pollutants in 34% of mechanically ventilated buildings (*Science Daily*, 8 February 1999).

Many people complain of asthma-like wheezing and tightness of the chest that improves on days away from the workplace. Headaches are another common problem. The pain associated with sick building syndrome is most often reported across the forehead, above the eyes and also at the back of the neck. The headache, like the lethargy which can come with it, usually gets worse as the day progresses and then improves once the person has left the building. Working all day under fluorescent lighting or looking at a computer screen could be the culprits.

Lethargy may be experienced by about 50% of workers in buildings that are air-conditioned, although about 15% of people working in naturally ventilated buildings also complained of this symptom in the same study (Finnegan *et al.*, 1984). Suggested causes include various 'indoor pollutants' such as ozone, carbon monoxide and formaldehyde from furniture and furnishings.

In a study of 4373 people working in 46 buildings (Wilson and Hedge, 1987), 80% reported ill health associated with being in their place of work. Of those, 29% said they had five or more symptoms. Lethargy was the most common complaint, followed by stuffy nose, dry throat, headache, itchy or dry eyes, runny nose, flu-like symptoms, difficulty in breathing and chest tightness.

People with clerical/secretarial jobs have more symptoms than their managers. This could be because clerical workers are often in open-plan offices where they have little control over their environment. As well as individual suffering, the cost to business can be high. A Swedish study (Edvardsson *et al.*, 2008) said that 45% of 'sick building syndrome' victims treated at hospital clinics no longer had the capacity to work.

12.4.2 Myalgic Encephalomyletis (ME) or Chronic Fatigue Syndrome (CFS)

The classic symptoms of ME or CFS bear some striking similarities with those reported in connection with air travel. The conditions are notoriously difficult to diagnose, but both usually have a clear starting point, often a bout of illness, which is why they are sometimes referred to as post-viral fatigue.

The most commonly reported symptoms include muscle and joint pain, extreme fatigue after exertion, forgetfulness, difficulty concentrating, disturbed sleep and flu-like symptoms. Some people also report heart palpitations, feeling faint, painful neck glands, a sore throat, headaches, nausea, skin sensitivity, digestive problems, mood swings and depression.

Anyone affected has an individual mix of symptoms, but added together they can have an intolerable impact on life. Many people find it impossible to work or study, but one strategy is for sufferers to 'save up' energy to enable them to do things by knowing exactly how much they can do and what the subsequent impact will be.

In 2008 a team led by prominent ME/CFS researcher Dr Jonathan Kerr published the results of a genetic study which identified seven different subtypes (BBC News, 5 May 2008):

- **Type 1** – high levels of depression and anxiety as well as poor sleep and high degrees of pain.
- **Type 2** – severe post-exertion fatigue, joint and muscle pains.
- **Type 3** – mildest form of the disease.
- **Type 4** – moderate levels of body pain and sleep problems.
- **Type 5** – most severe muscle weakness and predominance of gut problems.
- **Type 6** – associated with significant fatigue.
- **Type 7** – most severe form with high levels of pain, swollen glands and headaches.

Types 4 and 6 were the most common forms of the condition.

12.4.3 WiFi Sensitivity

Whether you want it or not, WiFi is becoming something that is hard to get away from, particularly in cities, but there are concerns that exposure to WiFi makes some people ill. WiFi has been banned in some public buildings in the US, UK and France, with detractors comparing its pervasive effects with 'second-hand smoke'.

Worries about the impact on health have not stopped places like Swindon from forging ahead and proclaiming itself the UK's first town to provide free WiFi internet access to all its residents, with 1400 secure internet access points. In the UK an estimated 80% of secondary and 50% of primary schools are thought to have WiFi installed in the classroom.

Those opposed to universal WiFi say that exposure can cause headaches, nausea, stomach upset, tinnitus and short-term memory loss. Sound familiar? Most of those symptoms can be found on the list of problems attributed to ill health caused by air travel, as well as those attributed to sick buildings, phone masts, electricity pylons and crop spraying.

Most experts say that WiFi should be as safe as other sources of http://www.osha.gov/SLTC/radiation_nonionizing/index.html non-ionising radiation such as cell phone networks and television broadcasts, but because it is a relatively new technology, no one can say what the medium- or long-term risks to health are likely to be. Several new studies are underway to see if some individuals do have a peculiar sensitivity to electro-magnetic fields, but hard evidence is thin on the ground.

As WiFi equipment tends to emit less intense radiation than other common sources, such as cell phones, most industry experts say there is no need for panic, but that has not stopped some calls for a more cautious approach.

Gone are the days when airlines prohibited personal electronic devices. Commercial operators are now rushing to install WiFi capability on their aircraft (it started to emerge as far back as 2005) and it seems inevitable that passengers will increasingly demand this connectivity. Those who are convinced that WiFi is detrimental to their health may soon be hard pressed to find a flight without it.

The use of a radio network (Airwave) by UK police forces has led to police officers preparing to sue their force over a series of illnesses they claim were caused by the use of the radio system on patrol. Scores of officers have claimed that radiation emissions from the system have caused symptoms which include nausea, headaches, stomach pains and skin rashes (*Telegraph*, 2 January 2010).

12.4.4 Electrical Power Line Sensitivity

Electro-sensitivity has made headlines in recent years, but is illness caused by an allergic-type reaction to power lines and invisible sources of electric and magnetic field radiation fact or fiction?

Some people believe that electrical and magnetic hypersensitivity impacts on their overall quality of life. Overhead power cables, substations, electrical wiring in the home and even cars have been blamed for a variety of health problems.

Comparisons have inevitably been drawn with sick building syndrome and symptoms of electrical and magnetic hypersensitivity include sleep disorders, tinnitus, depression, memory loss, headaches, joint pain and flu-like illness. More seriously, the condition has been linked to multiple sclerosis and some forms of cancer. The reported incidence is slightly higher in women than in men, and most sufferers report day-to-day variations in their sensitivity.

While claims of hypersensitivity have sometimes been met with cynicism (and several studies say they cannot find a proven link between illness and sources of electricity), the WHO established an international project in 1996 to assess the scientific evidence of possible health effects of exposure.

Studies into electrical hypersensitivity are difficult because of problems in accurately measuring dose and patterns of exposure, but research into the incidence of childhood leukaemia and some forms of motor neurone disease does highlight possible links with proximity to high-voltage power lines.

A seven-year study by the Childhood Cancer Research Group at Oxford University (www.ccrg.ox.ac) involved thousands of children and looked at the prevalence of high-voltage power cables near their homes. It found that those born or living near the power lines were 1.7 times more likely to contract leukaemia than those in a control group.

From the results of epidemiological investigations of 50 Hz fields from power lines, there remain concerns about a possible increased risk of childhood leukaemia associated with exposure to magnetic fields above about $0.4\,\mu T$. In this regard, it is important to consider the possible need for further precautionary measures.

12.4.5 New Car Syndrome

Some people like the smell of a new car so much that several years ago at least one materials company made a synthetic version, bottled it and sold it as air freshener. But some people believe that same 'smell' can make you ill.

A combination of substances used in car manufacture and which give the new interior its distinctive odour has been blamed for a variety of illnesses. The list of substances which could potentially impact on health is long, though most are also found in homes and offices on a daily basis. For example, the leather and fabrics in the car may be treated with chemicals such as bromides and antimony, used as flame retardants. Bromides are known to be harmful and have been linked to impaired memory.

Older cars whose 'new car' smell faded long ago have also been blamed for sickness. 'Sick' car syndrome caused by fumes from VOCs in car interior materials has been linked to sneezing, wheezing, chest complaints, nausea, drowsiness and eye irritation.

A study in the US (Air Sept, 1997) showed that car air-conditioning systems – both old and new – incubate mould and bacteria, which enter the passenger compartment when the system is switched on. The result can be symptoms similar to those cited in sick building syndrome and aerotoxicity, and, again, people with existing respiratory problems may be more at risk.

One commentator has called the list of toxins found in a car interior a 'chemical soup' and another has compared the inhalation of car toxins to glue sniffing (*Sydney Morning Herald*, 16 October 2009; www.foxnews.com, 9 April 2007; www.cnn.com, 24 October 2008).

The good news is that after the first three months of a car's life, the level of toxins reduces dramatically. Some manufacturers are tackling the problem by switching to natural resins, organic-based textiles and alternative leather treatments.

But a word of caution: at least one study (*Science Daily*, 10 April 2007) has found no evidence that 'new car smell' has any harmful effects at all. Researchers concluded that if you do not like the interior car smell, just roll down the window – not so easily done in an aircraft.

12.4.6 Household and Industrial Cleaning Products

Commercial airline operators have big cleaning bills, relying on an army of airport-based staff to clear up the mess left by passengers. It goes without saying that paying passengers expect their aircraft to be clean, hygienic, litter-free and smelling fresh when they board, but fumes and residue from cleaning products can cause problems. In order to avoid complaints about dirty aircraft, cleaners may well over-enthusiastically apply cleaning materials.

Once again, the main culprit is the VOCs which can result in eye, nose and throat irritation, headaches, chest complaints, loss of co-ordination and nausea. VOCs are widely used in household cleaning products and the frequent use of aerosols and air fresheners in the home has been linked with illness. An Australian study (BBC News, 25 August 2004) has suggested that exposure to fumes from cleaning products could cause asthma in children and increased incidence of stomach upsets and headaches.

Anyone with respiratory diseases may react to exposure at fairly low levels and a few people seem to react badly to a large range of chemicals at low concentrations, a condition called 'multiple chemical sensitivity'.

Environmentally friendly and 'green' cleaning products are available, and could help those sensitive to VOCs, but airlines face the challenge of efficient and effective cleaning in short timescales, as well as pressures on cost. Upholstery, carpets, hard surfaces such as tray tables and arm rests, overhead lockers, windows, toilet and galley areas and the cockpit all need regular attention. Aircraft on turnaround tend to undergo a quick tidy and wipe down; overnight stops give the opportunity for a deeper clean.

Restricting the use of effective cleaning chemicals becomes a balancing act as increased sensitivity about viruses and bacterial illness heightens public awareness on the whole subject of cleanliness. To many, anything less than 'spotless' in bathrooms and food serving areas is deemed unacceptable. Some people are actually reassured by the smell of cleaning materials – it is an indication that the environment is safe.

12.4.7 Discussion

Whilst the conditions described above do not appear to be immediately relevant to air travel, there are some connections. The aircraft contains a lot of avionic equipment that radiates and receives electro-magnetic energy. Examples are radios, radar and navigation systems (Moir and Seabridge, 2006). The aircraft electrical generation system generates electric power at alternating frequencies and high voltages similar to domestic electrical systems and distributes it around the aircraft to various equipment, including the equipment in the passenger seat (Moir and Seabridge, 2008) albeit at different voltages (115/200 V, 400 Hz rather than 240 V, 50 Hz).

The cabin is manufactured from materials and adhesives similar to those used in the car industry, and may contain VOCs. The aircraft is regularly cleaned and disinfected using industrial fluids which may also contain VOCs. It is most likely that every person in the cabin will be subject to exposure to these conditions, and may exhibit some symptoms.

The symptoms reported in a number of sources as a result of exposure to these topics are shown in Table 12.1 compared with the symptoms recorded by in-flight sufferers. Table 12.2 shows those symptoms which were reported as being unique to a condition. A few symptoms are 'catch-alls', such as 'flu', and may actually refer to many of the symptoms listed elsewhere in the table.

12.5 Serious Conditions

A number of serious conditions are declared in relation to some of the illnesses discussed in this book. These include cancer, kidney failure and neuromuscular diseases. These conditions may only arise many years after the initial contact with the potential stimulus and long-term research is required before a sensible statement can be made.

12.6 Advice to Industry

It is recognised that industry will have gained a great deal of experience over the years of aircraft development. However, it is worth recording some of the issues that need to be addressed in relation to air travel.

Table 12.2 Symptoms exhibited by airline passengers compared with people exposed to external conditions – singular symptoms.

Symptom	Jet lag	Fear of flying	Air Quality	WiFi[1]	EM[2]	New car[3]	Cleaning Fluid[4]	Sick building[5]	ME[6]	No of matches
Low Motivation			X							
Metallic Taste			X							
Hand/feet swelling			X							
Behavioural problems				X						
Fainting				X						
Light sensitive				X						
Seizure disorder					X					
Liver damage						X				
Decline in serum cholinesterase						X				
Nasal congestion								X		
Ear ache								X		
Stress								X		
Itching								X		

Sources:
[1] Adams (2008), Beschizza (2006), Philips (2009), *South Wales Evening Post* (2009), Moore (2007), *The Times* (2009).
[2] Graham (2005), US Environmental Protection Agency (1994), Lab News (2009), Air Quality Sciences (2006), Sixwise.com (2005), Article Alley (2005).
[3] Diagnoseme.com (2009), The Environmental Illness Resource (2009), Natural Matters (2006), Chemical Free (2009), EMF Blues (2009).
[4] BUPA (2004), Sixwise.com (2005), *The Ecologist* (2009), *Guardian* (2004), Institute of Environmental Health (1999), Babyworld (www.babyworld.co.uk).
[5] WHO (1983), Environmental Illness Resource (2009), Safe Workers (2009), Medicinenet.com.

12.6.1 Processes and Procedures

Company design processes and procedures must recognise the issue of the impact of flying on human health. The process should ensure that all appropriate measures are taken to protect aircraft occupants. The process also needs to take account of the difficulty of obtaining a meaningful understanding of the issues and of human variability to ensure that a balanced design is obtained.

Reviews should be structured to ensure that the right questions are posed to challenge engineers on their knowledge of human factor issues and the impact on the design. A checklist should be prepared to ensure completeness and consistency at each stage of the design review.

12.6.2 Independent Medical Advice

The designers of all manned military aircraft projects should seek independent aero-medical advice to ensure that the crew escape systems and life support systems do not endanger the health or life of the aircrew. The intention is to seek the best independent advice available; that is, not from within the company or the technical organisation designing military life support systems.

Designers of commercial aircraft systems also should seek access to such expertise so that the requirements of customers and developments in medical research can be understood and applied to the aircraft design. This advice demands knowledge of engineering and human physiology, which is usually best provided by someone with medical qualifications in the field of aviation medicine or aviation physiology. A relationship with a university or medical school is often beneficial for obtaining independent advice and access to a body of knowledge.

12.6.3 Research

It is beneficial to maintain awareness of trends in traveller issues and to do research to record such trends and potential solutions. It will be helpful to have a process in place to instruct employees on how to conduct such research and also to demonstrate to external bodies that the company is committed to excellence.

The need to understand the commercial issues of product liability and duty of care should ensure that funding is readily available for such research, either within the organisation of in the academic field. Industry needs to document this research and the results to demonstrate that it is taking a responsible approach to understanding the issues and to amending its design as appropriate.

12.6.4 Seeing the 'Big Picture'

The aircraft designer must take a systems engineering view and see the big picture. An example of this is shown in Figure 12.1 in which the situation described in Chapter 3 is expanded – with the benefit of hindsight admittedly.

In this diagram it can be seen that the primary driver for changing the oil additive was to improve the performance of the engine with subsequent benefits in performance and engine life. This in turn would result in lower maintenance costs and shorter aircraft downtime, giving a positive impact on operations.

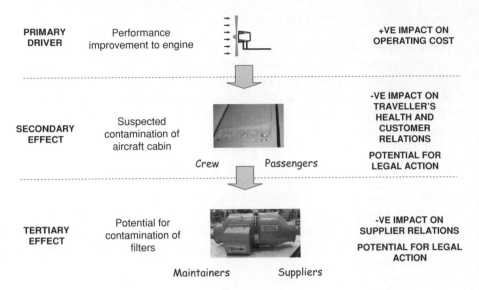

Figure 12.1   An expansion of the air system contamination situation in Chapter 3.

It has been alleged by the travelling public and crew members that their health has suffered as a result of suspected contamination in the cabin air system which is derived from the engine. It is further alleged that the onset of ill health coincides with the use of the new additive. This is a secondary effect that was not anticipated; it will have a negative impact on those people whose health has been damaged and may well lead to an erosion of public confidence and a potential for litigation raised by affected travellers.

If the contamination by organophosphates is proven, then a tertiary effect can be postulated which takes the impact to the realm of the equipment supply industry. Filters in the air system need to be changed and OBOGS beds need to be recharged at specified intervals. There now arises a potential for maintenance personnel to be exposed to contamination during this process. Note also that a similar technology is used to produce inert gas for fuel system fire suppression and the beds in this system will likewise be contaminated.

It is always worthwhile to examine the whole system when changes are being proposed to try to reduce the potential of similar effects. This is by no means a simple task and involves some real analysis at the conceptual stage of a new modification to understand the big picture, as well as a thorough design review process.

12.7 Advice to Operators

Aircraft fleet operators are the point in the system of interest at which there is close contact between the technical systems and the public or users. The operators have, therefore, a large part to play in reassuring the travelling public that their mode of transport is safe in that they will avoid accidents, and that it is also safe in the sense that it will not affect their health. This of course is equally applicable to their employees – the aircrew and cabin crew. It makes good business sense to keep passengers and employees aware at all times of the soundness of the aircraft fleet.

12.7.1 Promote Awareness

Operators should continue to make full use of in-flight magazines and web sites to give advice to travellers on how to enjoy their flight. Honesty may be the best policy here: admit to passengers that the airline operator is fully aware of the issues being published in the media and is offering advice on how to combat the conditions and explain what work is being done to find technical solutions.

12.7.2 Independent Medical Advice

Operators should seek independent medical advice to understand the issues in much the same way as the designers of aircraft. There may be opportunities to collaborate to obtain a thorough understanding of issues and to seek acceptable design solutions. This must be seen as collaboration and not collusion, hence the need for independence.

12.7.3 Research

Again, like industry, airline operators need to conduct their own research to understand what their customers are experiencing. There is a good case for independent research to collect and collate the experiences of all airline operators and a wide range of passenger experiences. This will go some way towards addressing the relatively small samples of data available.

12.8 Advice to Travellers

Travellers should make every effort to understand the risks to health and take measures to reduce their personal risk. Much information is available, and all airports and travel shops sell products that can make a difference to some conditions. If in any doubt at all, a doctor should be consulted.

12.8.1 Obtain Medical Advice

Before embarking on a trip that involves air travel it is always worthwhile to consult a general practitioner (GP), especially if there are any suspicions of personal or family history that may predispose the traveller to any condition. Business travellers should consult their company medical staff.

12.8.2 Improve Awareness

There is information available from many sources that provides good advice to travellers. Equally, there is information in the media, some of which is unfounded or based on anecdote. Travellers need to take a balanced view of such information.

A good place to start when planning a holiday or a trip is look at the airline operator's web site or in-flight magazine. There will be advice on some of the risks and what precautions to take in flight, although often not being explicit about why these precautions are required. Information on health is one thing that some budget operators have dropped as part of their bid to cut costs; but even if they do not have an in-flight magazine, most have information online.

The travel operator may also be a good source of advice.

As an example of good practice in the travel industry, the UK travel company Cox and Kings provides its customers with a leaflet written in conjunction with the Aviation Health Institute. The leaflet gives advice to travellers to help them to enjoy a good holiday. The contents are reproduced below (by permission of Cox and Kings):

HEALTHIER AND MORE COMFORTABLE AIR TRAVEL

Long haul travel affects the human body in various ways. The extent to which the different factors impact on us varies from person to person, however, there are a few simple steps that can be taken to lessen some of the discomforting symptoms that you may encounter.

The following advice has been prepared in conjunction with the Aviation Health Institute, a medical research charity that promotes health and well-being for air passengers. The following points are simply suggestions and you should always consult your G.P. with regard to any health concerns you may have prior to travel or before taking any medication.

10 GOOD IDEAS FOR BEFORE, DURING AND AFTER A FLIGHT

1. *Pre-flight exercise*: Aerobic exercises such as jogging, cycling and swimming, even for only half an hour, can help improve circulation for several hours afterwards and is recommended before a long flight.

2. *What to drink*: The air on aeroplanes is extremely dry and it is easy to become dehydrated. The best advice when flying is to drink as much water as possible – most airlines will provide drinking water on request, but it would be advisable to take a large bottle of mineral water on board with you. Alternatively, electrolytic drinks (available from most chemists) are highly effective, as they not only provide fluid but research shows that they help you retain it longer. Fizzy drinks are not recommended because the gas expands in the stomach at altitude. Alcoholic drinks are best avoided because they act as a diuretic and they are more potent when flying. If you do have an alcoholic drink it is always best to drink water afterwards.

3. *What to eat*: Many foods are gas-forming and, like fizzy drinks, can make the stomach swell uncomfortably. The best foods are fruits and salads, whilst the following should be <u>avoided</u> where possible: meat (which is difficult to digest), beans, peas, cabbage, cauliflower, cucumber, turnips, Brussels sprouts and anything with a high roughage content. In addition, if you find air travel stressful, it is best to avoid chocolates, soft cheese, citrus fruits, yeast extract and red wine, all of which can cause hypertension.

4. *Feeling fresh on arrival*: In order to lessen the debilitating effects of tiredness on arrival it is advisable to keep up your levels of beta-carotene. Scientific research has shown that drinking carrot juice two or three days before flying can make you feel much fresher on arrival by helping to retain more oxygen in the blood stream.

5. *Avoiding bugs*: The air on aeroplanes is filtered and recycled. Despite the filter, there will be some bugs in the cabin and because of the multinational nature of airline passengers, you may have a low resistance to some of these bugs. A dose of up to 100mg of Vitamin C prior to travelling and a dab of tea tree oil (available in most chemists) just below the nose during a flight can help prevent infection. Safer still is to use a 'Bugstoppers' face mask.

6. *Minimising ear pain*: Change in cabin pressure during take-off and landing may cause some discomfort in the ears. This is relieved by swallowing which releases pressure in the middle ear. Chewing gum and sucking sweets is recommended.

7. *What to wear*: The two factors to consider are determined by the cabin environment: your body inflates during flight and the cabin temperature can

be very cool. Therefore loose clothes, loose shoes and an extra sweater are a must. Tight and restrictive under garments and clothes should be avoided. Compression hose helps to improve circulation.

8. *Avoiding air sickness*: Passengers who are predisposed to air-sickness or anxiety may prefer a window seat. It is suggested that these conditions can be caused by disorientation due to the aircraft's movement and being able to see out of the window can help to alleviate the symptoms. Anti-sickness tablets and wrist-bands are available at any high street chemist.

9. *Swollen feet*: The lack of opportunities to move about during flight can result in the most common symptom of air travel: swollen feet, ankles and legs. This condition is due to poor circulation and here are several simple methods to encourage good circulation while on board: if possible walk about occasionally during the flight, or simulate walking by moving your feet up and down (ideally for 15 minutes of every hour); keep your hand luggage on the floor in front of you and rest your feet on it so that your thighs are clear of the edge of the seat; take an aspirin the day before the flight (subject to your doctor's advice). Passengers with a history of varicose veins or venous thrombosis should wear compression hose.

10. *On arrival*: To lessen the impact of jet lag you should synchronise yourself to local time as much as possible, limiting yourself to short naps during the daytime. Exposure to daylight on arrival helps speed up the process as sticking to local meal times. The best foods to choose on arrival are complex carbohydrates such as muesli bars, sandwiches, pasta, rice and potatoes, plus bio yoghurts and supplements of Vitamins E & C. Kiwi fruit can help alleviate constipation. Jet lag is normally more pronounced when flying from west to east.

The BBC Health web site (http://www.bbc.co.uk/health/physical_health) also provides information to travellers who suffer from travel sickness:

- Avoid heavy meals and alcohol before travelling.
- Keep still and close the eyes.
- Take anti-sickness medicines from the pharmacist BEFORE travel so they have time to be absorbed by the body.
- Take ginger or peppermint remedies. Ginger can be taken as a biscuit, tea or in crystallised form, while peppermint can be sucked as a sweet or taken as a tea. Ginger has been used by NASA to help astronauts combat nausea.
- Use acupressure, applied using a wristband or by pressing a finger against the middle of the inner wrist

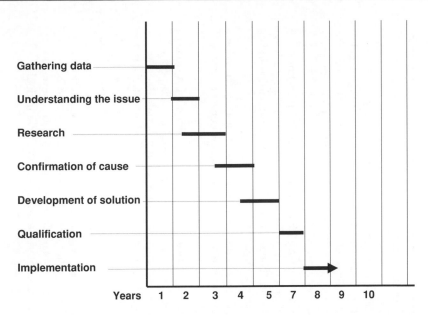

Figure 12.2 Some indicative timescales for the development of technical solutions.

12.9 What Can Be Done?

For many of the conditions described in this book, there can be no technical solutions designed into the aircraft or its systems. This is because it is physically impossible to enjoy fast, high-flying travel and to avoid some of the effects of the environment. For other conditions a technical solution is feasible, provided that those conditions can be shown to pose a significant health risk. However, in many cases the evidence to support the initiation of a technical investigation is low.

It takes time to develop a new technical solution even when there are clearly defined requirements. If the requirement is unknown or uncertain then considerable research and a feasibility study will need to be undertaken before a system solution can be designed, tested and qualified for use. Figure 12.2 shows the steps involved, and illustrates that even with only 12 months to progress each step, it may be more than five years before a feasible solution is installed.

What is needed to promote this process is good communication. Discussion on reported symptoms needs to be related to current research and where sufficient justification can be demonstrated then research topics can be

Table 12.3 Summary of conditions and recommended actions.

Condition	Education	Training	Awareness	Procedural	Design	Recommendation
General discomfort	x					Be aware of what a trip entails, relax – 'go with the flow' to avoid undue stress, rest frequently, eat and drink in moderation. Read airline and travel company advice on well-being. Take all opportunities for rest and relaxation
Jet lag	x					Synchronise with local time zones and meal times, sleep during flight, avoid alcohol, stimulants such as tea and coffee and too much food. Read airline and travel company advice
Fear of flying		x	x			Airlines to promote awareness of the issue. Traveller to take advantage of training and awareness courses provided by operators
Air quality					x	Maintain awareness of legislation, monitor customer feedback, work closely with users and incorporated design changes as and when appropriate. Monitor technology and changes in propulsion systems and fuels
Deep vein thrombosis	x				x	Travellers need to be aware of the issue, consult their doctor, take drugs only under medical supervision, exercise in flight, use support stockings cautiously. Read airline and travel company advice on well-being. Design to make room available for exercise and to consider novel seat shapes and configurations

Noise	x	x	Design to reduce noise at the ears, travellers to make use of earplugs or noise-cancelling headsets, avoid excessive noise environment before and after flight. Limit flight hours to achieve dose levels
Vibration	x	x	Design to reduce excessive vibration. Limit flying hours if this cannot be achieved. Monitor user observations
Cosmic radiation	x		Users to monitor flight hours above 26 000 t and to schedule crew rosters or apply flying hour limitations to maintain safe dose levels. Make use of suitable models and adjust for solar activity. Install radiation detectors
Non-ionising radiation	x	x	Design to prevent accidental radiation conditions, put warnings in flight manuals, schedule mode of radar to avoid accidental irradiation

(Continued)

Table 12.3 (Continued)

Condition	Education	Training	Awareness	Procedural	Design	Recommendation
Back pain	x				x	Design for comfortable seating and posture, make use of lumbar support and cushions, avoid work or leisure activities likely to inflame back problems before flying
Neck pain		x			x	Exercise routine as recommended by medical staff, maintain good posture. Design seats and helmet to reduce mass of head
Acceleration		x			x	Use AEA that limits exposure to g, use suitable straining manoeuvres, train on centrifuge or in flight
Decompression sickness	x					Avoid leisure pursuits that involve diving before flying. Avoid rapid changes in altitude/cabin pressure
Workstation	x			x	x	Ergonomic design, support for wrist and arms, good screen legibility and brightness control. User to be aware of issues and take breaks, change position and horizon

initiated, bearing in mind that funding will have to be sought. There are some examples of good communication in in-flight magazines, on web sites and in holiday company leaflets; however, they do not tell the whole story and they are one-way. There does not appear to be a mechanism for encouraging passengers and crew to open up a debate on their experiences. It must also be noted that in-flight magazines are read by passengers already flying, and they may not readily recognise the appropriate page, whilst web sites are available to those people with online access and time to search.

Table 12.3 summarises the major issues and indicates what can be done to address them.

Figure 12.3 Have you got everything?

12.10 Finally

Despite the horror stories, millions of people fly billions of miles every year and live to tell the tale. Some people do it because they have to, as it is part of their job, to some people it *is* their job, and to others it is a mechanism for going to their holiday destination or visiting friends and family. Many people enjoy it, most put up with it and some detest it. There is danger in every mode of travel. The airline industry has a relatively good accident rate, but it has somehow gained a reputation for damaging health. What is important is for everyone involved to be aware of the risks and to mitigate them where possible. Fasten your seat belts, enjoy the flight and remember to pack well (Figure 12.3)!

References

Action for ME (2003) November.

Adams, L. (2008) Electrosensitivity. *Scottish Daily Record*, 8 September.

Air Quality Sciences (2006).

Air Sept Inc. (1997) Customer Alert: Sick Car Syndrome Potentially Hazardous to Your Health, 26 June.

Article Alley (2005) 23 December.

Babyworld, www.babyworld.co.uk (accessed May 2010).

BBC (2004) 30 October.

BBC (2005) 3 November.

Beschizza, R. (2006) Wi-Fi as a Health Hazard. *Wired*, 12 December.

Bup (2008) January.

BUPA (2004) 28 October.

Chemical Free (2009).

Daily Mail (2004) 26 August.

Diagnoseme.com (accessed 27 April 2009).

Edvardsson, B., Stenberg, B., Bergdahl, J., Eriksson, N., Linden, G. and Widman, L. (2008) Medical and social prognoses of non-specific building-related symptoms (Sick Building Syndrome): a follow-up study of patients previously referred to hospital. *International Archives of Occupational and Environmental Health*, **81** (7), 805–512.

EMF Blues (2009).

Environmental Illness Resource (2009) 8 July.

Finnegan, M.J., Pickering, C.A. and Burge, P.S. (1984) The sick building syndrome: prevalence studies. *British Medical Journal (Clinical Research Edition)*, **289**, 1573–1575.

Graham, M.R. (2005) New Car Smell, treehugger.com (accessed 27 September 2005).

Green Cleaning Strategies, www.facilities.net (accessed May 2010).

Guardian (2004) 26 August.

Institute for Environment and Health (1999).

Lab News (2009).

ME Help Info.

Medicinenet.com (accessed May 2010).

Moir, I. and Seabridge, A. (2006) *Civil Avionic Systems*, John Wiley & Sons, Ltd.

Moir, I. and Seabridge, A. (2008) *Aircraft Systems*, John Wiley & Sons, Ltd.

Moore, V. (2007) *Daily Mail*, 27 April. http://www.dailymail.co.uk/news/article-449679/power_lines_link_cancer_new_alert.html (accessed Date Month Year).

Natural Matters (2006) February.

Nbc10.com (accessed May 2010).

Philips, A. (2009) Is Electric Smog Causing Your Headache? Daily Mail on-line, 19 November.

Safe Workers (2009).

Science Daily (2008) 19 May.

Scottish Parliament Public Petitions Committee, Consideration PE812.

Sixwise.com (accessed 14 December 2005).

South Wales Evening Post (2009) 18 March.

theautochannel.com (accessed 26 June 1997).

The Ecologist (2009) 3 March.

The Environmental Illness Resource (2009) 28 July.

The Times (2009) 19 November.

US Environmental Protection Agency (1994).

WHO (1983).

Further Reading

Farrow, A., Taylor, H., Northstone, K. *et al.* (2003) ALSPAC Study Team. Symptoms of mothers and infants related to volatile organic compounds in household products. *Archives of Environmental Health*, **58**, 633–641.

Institute for Environment and Health (1999) Volatile organic compounds (including formaldehyde) in the home, www.le.ac.uk/ieh (accessed February 2010).

Levitt, B.B. (1995) *Electromagnetic Fields: A Consumer's Guide to the Issues and How to Protect Ourselves*, Harvest Original.

London Hazards Centre (1990) Sick Building Syndrome – causes, effects and control, www.lhc.org.uk (accessed May 2010).

Rumchev, K., Spickett, J., Bulsara, M *et al.* (2004) Association of domestic exposure to volatile organic compounds with asthma in young children. *Thorax*, **59**, 729–730.

Useful Web Sites

Powerwatch.org.uk

www.who.int

Bibliography

We have made use of a wide range of sources during the preparation of this book. Our search has ranged across newspapers and magazines, books, published papers, personal contacts, interviews, TV and radio and, of course, the internet. We have done our best to extract the most relevant information to support the book. For those wishing to follow our footsteps, there is a lot of duplication of information, but if you want to follow up a particular line of research or examine issues in greater depth, then there is plenty of material available.

AAIB Bulletin No. 7/2005: EW/G2004/11/08 & EW/G2004/11/12: Boeing 757-236, G-BPEE. 12, 16 and 23 November 2004.

Abou-Donia, Mohamed B. (2009) Chemical-Induced Brain Injury. Presentation given to GCAQE, London, http://www.gcaqe.org (presentations), (accessed February 2010).

Adams, L. (2008) Electrosensitivity. *Scottish Daily Record*, 8 September.

Agampodi, S.B., Dharmaratne, S.D. and Agampodi, T.C. (2009) Incidence and predictors of on-board injuries among Sri Lankan flight attendants. *BMC Public Health*, **9**, 227.

Air Sept Inc. (1997) Customer Alert: Sick Car Syndrome Potentially Hazardous to Your Health, **26** June.

Albano, J.J. and Stanford, J.B. (1998) Prevention of minor neck injuries in F-16 pilots. *Aviation, Space and Environmental Medicine*, **69**, 1193–1199.

Alves-Pereira, M. and Castelo Branco, N. (1999) Vibroacoustic Disease: the Need for a New Attitude Towards Noise, Center for Human Performance, Portugal, www.lowertheboom.org (accessed February 2010).

Australian Department of Defence (2005) Organophosphate and amine contamination of cockpit air in the Hawk, F-111 and Hercules C-130 Aircraft, DSTO publications, October.

BAE Systems verbal evidence to Australian Senate Inquiry, 2000.

Bagshaw, M. (2001) Traveller's thrombosis – a review of Deep Vein Thrombosis associated with air travel. Air Transport Medicine Committee. *Aerospace Medical Association*, **72**, 848–851.

Ballard, T. *et al.* (2000) Cancer incidence and mortality among flight personnel: a meta analysis. *Aviation Space Environmental Medicine*, **71** (3), 216–224.

Barish, R. (1996) *The Invisible Passenger: Radiation Risks for People Who Fly*, Advanced Medical Publishing.

Barish, R.J. (1999) In-flight radiation – counseling patients about risk. *Journal of the American Board of Family Practice*, **12**, 195–199.

Barnett-Jones, F. (2008) *Tarnish 6: The Biography of Test Pilot Jimmy Dell*, Old Forge Publishing.

Beighton, P.H. and Richards, P.R. (1968) Cardiovascular disease in air travelers. *British Heart Journal*, **30**, 367–372.

Belcaro, G., Cesarone, M.R., Nicolaides, A.N. *et al.* (2003) The LONFLIT4-VENORUTON study: a randomized trial prophylaxis of flight-edema in normal subjects. *Clinical and Applied Thrombosis/Hemostasis*, **9** (1), 19–23.

Belcaro, G., Cesarone, M.R., Rohdewald, P. *et al.* (2004) Prevention of venous thrombosis and thrombophlebitis in long-haul flights with Pycnogenol. *Clinical and Applied Thrombosis/Hemostasis*, **10** (4), 373–377.

Belobaba, P., Odoni, A. and Barnhadt, C. (eds) (2009) *The Global Airline Industry*, John Wiley & Sons, Ltd.

Bendz, B., Rostrup, M., Sevre, K. *et al.* (2000) Association between acute hypobaric hypoxia and activation of coagulation in human beings. *The Lancet*, **356**, 1657–1658.

Beschizza, R. (2006) Wi-Fi as a Health Hazard. *Wired*, 12 December.

Blettner, M., Zeeb, H., Auvinen, A. *et al.* (2003) Mortality from cancer and other causes among male airline cabin attendants in Europe: a collaborative cohort study in eight countries. *American Journal of Epidemiology*, **158** (1), 35–46.

Blettner, M., Zeeb, H., Auvinen, A. *et al.* (2003) Mortality from cancer and other causes among male airline cockpit crew. *International Journal of Cancer*, **106** (6), 946–953.

Bridger, R.S., Groom, M.R., Jones, H. *et al.* (2002) Task and postural factors are related to back pain in helicopter pilots. *Aviation, Space and Environmental Medicine*, **73** (8), 805–811.

Bryson, B. (2003) *A Short History of Nearly Everything*, Doubleday.

Burn, L., Sinel, M.S. and Deardoff, W.W. (2007) *Treating Your Back and Neck Pain for Dummies*, John Wiley & Sons, Ltd.

Burström, L., Lindberg, L. and Lindgren, T. (2006) Cabin attendants' exposure to vibration and shocks during landing. *Journal of Sound and Vibration*, **298** (3), 601–685.

CASA (2007) Air Safety & Cabin Air Quality, Jim Coyne, A/g General Manager Manufacturing, Certification & New Technologies Office, Presentation.

CBC News Online (2006) Indepth: Health – The perils of traveller's thrombosis. 11 January.

Chen, J. and Mares, V. (2008) Estimate of the dose to the fetus during commercial flights. *Health Physics*, **95** (4), 407–412 .

Chodick, G., Bekiroglu, N., Hauptmann, M. *et al.* (2008) Risk of cataract after exposure to low doses of ionizing radiation: a 20-year prospective cohort study among US radiologic technologists. *American Journal of Epidemiology*, **168**, 620–631.

Clarke, M. (2003) Editorial – Cosmic exposures. HPA e-bulletin No. 4.

Commonwealth of Australia Senate Hansard, Monday, 13 August 2007 and Thursday, 20 September 2007, Aircraft Cabin Air Quality – Senator O'Brien.

Contaminated Air Protection Conference Proceedings (2005) Imperial College, London, 20–21 April.

Cucinotta, F.A., Manuel, F.K., Jones, J. *et al.* (2001) Space radiation and cataracts in astronauts. *Radiation Research*, **156** (Pt 2), 460–466.

Cummin, A. and Nicholson, A. (eds) (2002) *Aviation Medicine and the Airline Passenger*, Arnold.

Cummin, A.R.C. and Nicholson, A.N. (2002) The cabin environment, in *Aviation Medicine and the Airline Passenger* (eds A. Cummin and A. Nicholson), Arnold.

Davies, J.R., Johnson, R., Stepanek, J. and Fogart, J.A. (2008) *Fundamentals of Aerospace Medicine*, 4th edn, Walters Kluwer.

Davis, R.L. and Mostofi, F.K. (1993) Cluster of testicular cancer in police officers exposed to hand-held radar. *American Journal of Industrial Medicine*, **24** (2), 231–233.

Delvin, D. (2009) *Backache – What You Need to Know*, Sheldon Press.

Diski, J. (2006) *On Trying to Keep Still*, Little, Brown.

Edvardsson, B., Stenberg, B., Bergdahl, J., Eriksson, N., Linden, G. and Widman, L. (2008) Medical and social prognoses of non-specific building-related symptoms (Sick Building Syndrome): a follow-up study of patients previously referred to hospital. *International Archives of Occupational and Environmental Health*, **81** (7), 805–512.

Environmental Protection Agency (2006) EPA Green Book, www.epa.gov/air (accessed February 2010).

European Standard EN 1822-1. Draft. High Efficiency Air Filters (EPA, HEPA, ULPA).

Farrow, A., Taylor, H., Northstone, K. *et al.* (2003) ALSPAC Study Team. Symptoms of mothers and infants related to volatile organic compounds in household products. *Archives of Environmental Health*, **58**, 633–641.

Finnegan, M.J., Pickering, C.A. and Burge, P.S. **(1984)** The sick building syndrome: prevalence studies. *British Medical Journal (Clinical Research Edition)*, **289**, 1573–1575.

Flight Health. The problem of cosmic radiation. flight health.org/cosmic-radiation/ cosmic-radiation-the-problem (accessed February 2010).

Fortescue, P., Stark, J. and Swinerd, G. (2003) *Spacecraft Systems Engineering*, John Wiley & Sons, Ltd.

Freidburg, W., Copeland, K., Duke, F.E. *et al.* (2000) Radiation exposure during air travel: guidance provided by the Federal Aviation Administration for air carrier crews. *Health Physics*, **79**, 591–595.

Geeze, D.S. (1998) Pregnancy and in-flight cosmic radiation. *Aviation, Space and Environmental Medicine*, **69**, 1061–1064.

German Ministry of Transport, Secretary of State Ulrich Kasparick, Question to MP Winfried Hermann of Bundnis90/Greenparty in regards to contaminated cabin air on board civil airliners, printed matter 16/12023, 3 March 2009.

Graham, M.R. (2005) New Car Smell, treehugger.com (accessed 27 September 2009).

Graham-Cumming, A.N. (n.d.) Moulded lumbar supports for aircrew backache: comparison of effectiveness in fixed and rotary wing aircrew, Headquarters, Personnel and Training Command, RAF Innsworth, Gloucester. http://ftp.rta.nato.int/public/PubFullText/RTO/MP/RTO-MP-019/$MP-019-35.PDF (accessed 7 May 2010).

Gulf War Illness and the Health of Gulf War Veterans (2008) Scientific Findings and Recommendations –break; Research Advisory Committee on Gulf War Veterans Illnesses – US Department of Veterans Affairs, Washington, DC.

Gundestrup, M. and Storm, H.H. (1999) Radiation-induced acute myeloid leukaemia and other cancers in commercial jet cockpit crew: a population-based cohort study. *The Lancet*, **354**, 2029.

Gundestrup, M., Andersen, M.K., Sveinbjornsdottir, E. *et al.* (2000) Cytogenetics of myelodysplasia and acute myeloid leukaemia in aircrew and people treated with radiotherapy. *The Lancet*, **356**, 2158.

Haldorsen, T., Reitan, J.B. and Tuelen, U. (2000) Cancer incidence among Norwegian airline pilots. *Scandinavian Journal of Work, Environment and Health*, **26**, 106–111.

Hamada, K, Doi, T., Sakurai, M. *et al.* (2002) Effects of hydration on fluid balance and lower-extremity blood viscosity during long airplane flights. *Journal of the American Medical Association*, **287**, 844–845.

Hammar, N. *et al.* (2002) Cancer incidence in airline and military pilots in Sweden 1991–1996. *Aviation, Space and Environmental Medicine*, **73** (1), 2–7.

Hansen, O.B. and Wagstaff, A.S. (2001) Low back pain in Norwegian helicopter aircrew. *Aviation, Space and Environmental Medicine*, **72**, 161–164.

High Life, British Airways in-flight magazine.

Hocking, M.B. and Hocking, D. (2005) *Air Quality in Airline Cabins and Related Enclosed Spaces*, Springer.

Holland Herald, KLM in-flight magazine.

House of Lords Select Committee on Science and Technology, 5th Report (15 November 2000) Air Travel and Health, Chapter 1: Summary and Recommendations.

Hughes, R., Weatherall, M., Wilsher, M. and Beasley, R. (2004) Venous thromboembolism in long-distance air travellers. *The Lancet*, **363**, 896–897.

Hunter, R.M.C. (2002) Cosmic radiation, in *Aviation Medicine and the Airline Passenger* (eds A. Cummin and A. Nicholson), Arnold.

Hunter, R.M.C. (2003) *Protection of Aircrew from Cosmic Radiation: Guidance Material*, CAA.

Iles, R.H.H.A., Jones, J.B.L., Bentley, R.D. *et al.* (2003) The effects of solar particle events at aircraft altitudes. Proceedings of ESA Space Weather Workshop: Looking towards a European space weather programme (17–19 December 2001), pp. 121–124.

Institute for Environment and Health (1999) Volatile organic compounds (including formaldehyde) in the home, www.le.ac.uk/ieh (accessed February 2010).

Irvine, D and Davies, D.M. (1999) British Airways flight-deck mortality study. *Aviation, Space and Environmental Medicine*, **70** (6), 548–55.

Jonathan, D.A. (2002) Otolaryngology, in *Aviation Medicine and the Airline Passenger* (eds A. Cummin and A. Nicholson), Arnold.

Jukes, M.L. (2004) *Aircraft Display Systems*, John Wiley & Sons, Ltd.

Jones, J.A., Hart, S.F., Baskin, D.S., Effenhauser, R., Johnson, S.L., Novas, M.A., Jennings, R. and Davis, J. (2000) Human and behavioral factors contributing to spine-based neurological cockpit injuries in pilots of high-performance aircraft: recommendations for management and prevention. *Military Medicine*, **165** (1), 6–12.

Jukes, M.L. (2004) *Aircraft Display Systems*, John Wiley & Sons, Ltd.

Kelman, C.W., Kortt, M.A., Becker, N.G. *et al.* (2003) Deep vein thrombosis and air travel: record linkage study. *British Medical Journal*, **327**, 1027.

Kos, C.A. National Alliance for Thrombosis and Thromboplilia. stopetheclot.com (accessed February 2010).

Lawson, C. (2010) Environmental control systems, in *Wiley International Encyclopaedia of Aerospace*, Vol. **8**, Chapter 8.3.08 .

Lean, G. (2009) Mobiles and Cancer: The Plot Thickens. *Daily Telegraph*, 12 September.

Levitt, B.B. (1995) *Electromagnetic Fields: A Consumer's Guide to the Issues and How to Protect Ourselves*, Harvest Original.

Lim, M.K. and Bagshaw, M. (2002) Cosmic rays: are air crew at risk? *Occupational and Environmental Medicine*, **59**, 428–432.

Linnersjö, A., Hammar, N., Dammström, B.-G. *et al.* (2003) Cancer incidence in airline cabin crew: experience from Sweden. *Occupational and Environmental Medicine*, **60**, 810–814.

London Hazards Centre (1990) Sick Building Syndrome – causes, effects and control, www.lhc.org.uk (accessed May 2010).

Loomis, T.A., Hodgson, J.A., Hervig, L. and Prusacyck, W.K. (1999) *Neck and Back Pain in E2-C Hawkeye Aircrew*, Storming Media.

Lowden, A. and Akerstedt, T. (1998) Retaining sleep and wake patterns in aircrew on a 2-day layover on westward long distance flights. *Chronobiology International*, **15** (4), 365–376.

Mackenzie-Ross, S.J. (2008) Cognitive function following exposure to contaminated air on commercial aircraft. A case series of 27 airline pilots seen for clinical purposes. *Journal of Nutritional and Environmental Medicine*, **17** (2), 111–126.

Mackenzie-Ross, S.J., Harper, A.C. and Burdon, J. (2006) Ill health following reported exposure to contaminated air on commercial aircraft: psychosomatic disorder or neurological injury? *Journal of Occupational Health & Safety: Australia & New Zealand*, **22** (6), 521–528.

Mansfield, N.J. (2002) Proposed EU Physical Agents Directives on noise and vibration, in *Contemporary Ergonomics* (ed. P.T. McCabe), Taylor & Francis.

Mansfield N.J. (2005). *Human Response to Vibration*. CRC Press, Boca Raton. ISBN 0-415-28239-X.

Marais, K. and Waitz, I.A. (2009) Air Transport and the Environment, in *The Global Airline Industry* (eds P. Belobaba *et al.*), John Wiley & Sons, Ltd, pp. 405–436.

Marx, J. (2009) Altitude Induced Venous Failure, Bupa factsheet, June.

May, Captain Joyce (2005) Considerations Regarding Flying and Pregnancy. *Science Daily*, 1 September.

McKenas, D. (American Airlines) www.airhealth.com (accessed February 2010).

Mckeown, M. (2003) Research Report on Deep Vein Thrombosis in Air Travellers. *International Health News*, No. 142.

Michaelis, S. (2007) *Aviation Contaminated Air Reference Manual*. Susan Michaelis.

Michaelis, S. (2007) Letter from Captain Susan Michaelis to the 2007 UK House of Lords Inquiry, www.publications.parliament.uk (accessed February 2010).

Michaelis, S., Winder, C., Hooper, M. and Harper, A. (2008) Critique of the UK Committee on Toxicity Report on Exposure to Oil Contaminated Air on Commercial Aircraft and Pilot Ill Health, www.aopis.org/ScientificReports.html (accessed February 2010).

Midkiff, A.H., Hansman, R.J. and Reynolds, T.G. (2009) Airline Flight Operations, in *The Global Airline Industry* (eds P. Belobaba *et al.*), John Wiley & Sons, Ltd.

Mittermayr, M., Fries, D., Gruber, H. *et al.* (2007) Leg edema formation and venous blood low velocity during a simulated long-haul flight. *Thrombosis Research*, **120** (4), 497–504.

Miyagi, M. (2005) *Serious Accidents and Human Factors*, John Wiley & Sons, Ltd.

Mobil Oil Corporation (1983) *Mobil Jet Oil II*, Environmental Affairs and Toxicology Department, New York, Correspondence, available at www.exxonmobil.com (accessed February 2010).

Moir, I. and Seabridge, A. (2004) *Design and Development of Aircraft Systems*, John Wiley & Sons, Ltd.

Moir, I. and Seabridge, A. (2006) *Military Avionics Systems*, John Wiley & Sons, Ltd.

Moir, I. and Seabridge, A.G. (2008) *Aircraft Systems*, 3rd edn, John Wiley & Sons, Ltd.

Moore, V. (2007) *Daily Mail*, 27 April.

Morris, C.B., Wg Cdr (2010) RAF Aviation Medicine Training of RAF Aircrew. ftp.rta.nato.int/public/pubfulltext/rto/mp/rto-mp-021 (accessed May 2010) .

Nachenson, A. and Jonsson, E. (eds) (2000) *Neck and Back Pain: The Scientific Evidence of Causes, Diagnosis and Treatment*, Lipincott Williams and Wilkins.

New York Times, October 2000.

Nicholas, J.S., Lackland, D.T., Dosemechi, M. *et al.* (1998) Mortality among US commercial pilots and navigators. *Journal of Occupational & Environmental Medicine*, **40** (11), 980–985.

Nicholson, A.N. (2002) Sleep Disturbance and Jet Lag, in *Aviation Medicine and the Airline Passenger* (eds A. Cummin and A. Nicholson), Arnold.

NTP Chemical Repository data (Radian Corporation, 29 August 1991), Tricresyl phosphate.

Observer, Sunday 14 January 2001.

Office Ergonomics Training, www.office-ergo.com (accessed 11 May 2010).

Olas, B., Wachowicz, B., Saluk-Juszczuk, J. and Zieluski, T. (2002) Effect of reseveratol, a natural polyphenolic compound, on platelet activation induced by endotoxin or thrombin. *Thrombosis Research*, **7** (3–4), 141–145.

Parry, D. (2009) *Moon Shot: The Inside Story of Mankind's Greatest Adventure*, Ebury Press.

Parsi, K., McGrath, M.A. and Lord, R.S. (2001) Traveller's venous thromboembolism. *Cardiovascular Surgery*, **9** (2), 157–158 .

Petren-Mallmin, M. and Linder, J. (2001) Cervical spine degeneration in fighter pilots and controls: a 5-yr follow-up study. *Aviation, Space and Environmental Medicine*, **72** (5), 443–446.

Philips, A. (2009) Is Electric Smog Causing Your Headache? Daily Mail on-line, 19 November.

Pook, J. (2009) *Flying Freestyle: An RAF Fast Jet Pilot's Story*. Pen and Sword Books.

Pukkala, E. *et al.* (2003) Cancer incidence among 10,211 airline pilots: Nordic study. *Aviation, Space and Environmental Medicine*, **74**, 699–706.

Quick Reference Guide for Health Care Providers (1999) Health impact of exposure to contaminated supply air on commercial aircraft, Division of Occupational Health and Environmental Medicine, San Francisco.

Rafnsson, V., Hrafnkelsson, J. and Tulinius, H. (2000) Incidence of cancer among commercial airline pilots. *Occupational and Environmental Medicine*, **57**, 175–179.

Rafnsson, V., Hrafnkelsson, J., Tulinius, H. *et al.* (2003) Risk factors for cutaneous malignant melanoma among aircrews and a random sample of the population. *Occupational and Environmental Medicine*, **60**, 815–820.

Rafnsson, V., Sulem, P., Tulinius, H. and Hrafnkelsson, J. (2003) Breast cancer risk in airline cabin attendants: a nested case-control study in Iceland. *Occupational and Environmental Medicine*, **60**, 807–809.

Rafnsson, V., Olafsdottir, E., Hrafnkelsson, J. *et al.* (2005) Cosmic radiation increases the risk of nuclear cataract in airline pilots. A population-based case-control study. *Archives of Ophthalmology*, **123**, 1102–1105.

Rainford, Commodore D.J. and Gradwell, Gr Capt D.P. (2006) *Ernsting's Aviation Medicine*, Hodder Arnold.

Rayman, R.B. and McNaughton, G.B. (1983) Smoke/fumes in the cockpit. *Aviation, Space and Environmental Medicine*, **54**, 738–740.

Reyneke, D. (2001) *In-flight Fitness*, Orion Books Ltd.

Rolls Royce, Germany (2003) BRE Air Quality Conference, London.

Rumchev, K., Spickett, J., Bulsara, M *et al.* (2004) Association of domestic exposure to volatile organic compounds with asthma in young children. *Thorax*, **59**, 729–730.

SAE Aviation Information Report: **1539**, 30 January 1981.

Samel, A., Wegmann, H.M. and Vejvoda, M. (1997) Aircrew fatigue in long-haul operations. *Accident Analysis and Prevention*, **29** (4), 439–452.

Sargent, P., Lt MD and Bachmann, A., Lt MD (n.d.) Back pain in the naval rotary wing community, Naval Safety Center: http://safetycenter.navy.mil/Aviation/articles/back_pain.htm (accessed 7 May 2010).

Science Daily (2008)19 May.

Schoberberger, W., Fries, D., Mittermayr, M. *et al.* (2002) Changes of biochemical markers and functional tests for clot formation during long-haul flights. *Thrombosis Research*, **108** (1), 19–24.

Schoberberger, W., Mittermayr, M., Fries, D. *et al.* (2007) Changes in blood coagulation of arm and leg veins during a simulated long-haul flight. *Thrombosis Research*, **119** (3), 293–300.

Schreijer, A., Cannegieter, S.C., Doggen, C.J.M. and Rosendaal, F.R. (2008) The effect of flight-related behaviour on the risk of venous thrombosis after air travel. *British Journal of Haematology*, **144** (3), 425–429.

Scott, R., Hart, Y., Holdstock, D.J. and Lynn, W. (1985) Medical emergencies in the air. *The Lancet*, **325** (8424), 353–354.

Scott, W.B. (2004) Pulling gs on earth. *Aviation Week and Space Technology*, 5 January.

Scurr, J.H., Machin, S.J., Bailey-King, S. *et al.* (2001) Frequency and prevention of symptomless deep-vein thrombosis in long-haul flights: a randomised trial. *The Lancet*, **357**, 1485–1488.

Senate Rural & Regional Affairs & Transport References Committee. Air safety & cabin air quality in the BAe 146 aircraft, Parliament of Australia, Canberra, Final report, October 2000, Sections 5.31–5.32.

Sharine, A. (2009) A Scanner Darkly. *Guardian*, 15 October.

Sharma, S., Wg Cmdr and Upadhyay, A.D., Wg Cmdr (2000) Is backache a serious malady among Indian helicopter pilots and low backache among Chetak helicopter pilots? Trial of lumbar cushions at a flying unit. *Indian Journal of Aerospace Medicine*, **44**, 56–63.

Sigurdson, A.J. and Ron, E. (2004) Cosmic radiation exposure and cancer risk among flight crew. *Cancer Investigation*, **22** (5), 743–761.

Simon-Arndt, C.M., Yuan, H. and Hourani, L.L. (1997) Aircraft type diagnosed back disorders in US Navy pilots and aircrew. *Aviation, Space and Environmental Medicine*, **68**, 1012–1018.

Simpson, K. (1940) Shelter deaths from pulmonary embolism. *The Lancet*, **ii**, 744.

Smith, A. (2005) *Moon Dust*, Bloomsbury.

Snijders, C.J. (2010) Study of low back pain in crewmembers in space flight. International Space Station fact sheet, http://www.nasa.gov/mission_pages/station/science/experiments/Mus.html (accessed May 2010).

South, T. (2004) *Managing Noise and Vibration at Work – A Practical Guide to Assessment, Measurement and Control*, Elsevier.

Starmer-Smith, C. (2009a) 'Toxic' Cabin Air Found in New Plane Study. *Telegraph Travel*, 14 February.

Starmer-Smith, C. (2009b) Illness Among Cabin Crew Heightens Toxic Air Fears. *Telegraph Travel*, 18 July.

Stott, J.R. (2002) Airsickness, in *Aviation Medicine and the Airline Passenger* (eds A. Cummin and A. Nicholson), Arnold.

Stroud, R. (2009) *A Book of the Moon*, Doubleday.

Symington, I.S. and Stack, B.H. (1977) Pulmonary embolism after travel. *British Journal of Diseases of the Chest*, **71** (2), 138–140.

Taylor, G.C., Bentley, R.D., Conroy, T.J. *et al.* (2002) The evaluation and use of a portable TEPC systems for measuring in-flight exposure to cosmic radiation. *Radiation Protection Dosimetry*, **99** (1–4), 435–438.

The Anti-G Straining Manouevre: www.tpub.com/content/aviation2/P-868/P-8680094.htm (accessed 11 May 2010).

Tokumaru, O., Haruki, K. , Bacal, K. *et al.* (2006) Incidence of cancer among female flight attendants: a meta-analysis. *Journal of Travel Medicine*, **13** (3), 127–132.

Turner v. *Eastwest Airlines* [2009] NSWDDT, 5 May 2009, Australian Court.

UK COT Report (1999) Long term sequelae of acute poisoning. Committee on Toxicity of Chemicals in Food, Consumer Products and the Environment: Organophosphates: Executive Summary, Department of Health, London.

UK Government Hansard 66599, 4 February 1999, column 737.

UK HSE (1998) Organophosphates: HSE: MS17: Medical aspects of occupational exposures to organophosphates. Draft revision 23, November.

Walker, G. (2007) *An Ocean of Air: A Natural History of the Atmosphere*, Bloomsbury.

Whelan, E.A. (2003). Cancer incidence in airline cabin crew. *Occupational Environmental Medicine*, **10** (11), 805.

Yeoell, L. and Kneebone, R. (2003) On-Board Oxygen Generation Systems (OBOGS) for in-service military aircraft – the benefits and challenges of retro-fitting, Internal Honeywell report.

Yong, L.C., Sigurdson, A.J., Ward, E.M. *et al.* (2009) Increased frequency of chromosome translocations in airline pilots with long-term flying experience. *Occupational and Environmental Medicine*, **66**, 56–62.

Useful Web Sites

aerotoxic.org
af.mil
Aircrewhealth.com
airhealth.org
airsafe.org
antijetlagdiet.com
britishairways.com/travel/healthcosmic/public/en.gb
dft.gov.uk
encyclopedia.com
faa.gov/safety/programs_initiatives/aircraft_aviation/cabin_safety/rec_imp

fearfreeflying.co.uk
flightglobal.com
futureflite.com
hepafilter-pro.com
hps.org (Health Physics Society)
hse.gov.uk
medicine.net
nojetlag.com
nzherald.co.nz
ohrca.org/healthguide
patient.co.uk
publications.parliament.uk/pa/ld200708/ldselect/ldsctech
science.nasa.gov
solarstormwatch.com
spine-health.com
spineuniverse.com
stfc.ac.uk (Science & Technologies Facilities Council)
wddty.com (What doctors don't tell you)
who.int
who.int/ionizing_radiation/env/cosmic/WHO_Info_Sheet_Cosmic_
Radiation.pdf

Index

Air Travel and Health: A Systems Perspective   Allan Seabridge and Shirley Morgan
© 2010 John Wiley & Sons, Ltd